DEVELOPING STRATEGIES FOR THE MODERN INTERNATIONAL AIRPORT

Developing Strategies for the Modern International Airport
East Asia and Beyond

ALAN WILLIAMS
Massey University, New Zealand

ASHGATE

Published by
Ashgate Publishing Limited
Gower House
Croft Road
Aldershot
Hampshire GU11 3HR
England

Ashgate Publishing Company
Suite 420
101 Cherry Street
Burlington, VT 05401-4405
USA

Ashgate website: http://www.ashgate.com

British Library Cataloguing in Publication Data
Williams, Alan
    Developing strategies for the modern international airport
    : East Asia and beyond
    1.Airports - Economic aspects - East Asia
    I.Title
    387.7'36'095

Library of Congress Cataloging-in-Publication Data
Williams, Alan, 1934-
    Developing strategies for the modern international airport : East Asia and beyond / by Alan Williams.
        p. cm.
    Includes bibliographical references and index.
    ISBN 0-7546-4445-6
    1. Airports--Design and construction. I. Title.

    TL725.W55 2006
    387.7'36--dc22

2006000086

ISBN-10: 0 7546 4445 6

Printed and bound in Great Britain by Antony Rowe Ltd, Chippenham, Wiltshire.

Contents

List of Figures

List of Tables

Preface

This book may be described as an attempt to identify and analyse some of the complex issues and problems now being faced by modern international airports, which are destined to have a fundamental influence on their future strategic goals and operational functions. The need to respond in a proactive way to quite profound economic challenges driven by the internationalization of the global economy emerged during the latter part of the twentieth century and continues to grow apace. It is driven by the fact that the economic geography of world commerce and trade is itself changing direction, in a multitude of ways. These changes bring in their train the need to address the identification of the major future role of transportation systems. With structural shifts taking place in the demographics of urbanization, regionalization and new market development, urgent attention to such matters as congestion and environmental impacts, is also high on the international agenda. As a consequence the predominant emphasis throughout this book will be on the economic and geopolitical influences that are currently shaping the role of the international airport, both nationally and in the much larger contexts of urbanization and globalization.

It is also very important that airport managements are made aware, and are able to respond in an appropriate and strategic way, to the fact that there are strong pressures being exerted by geopolitical, multinational business and environmental constituencies. They, in turn, tend to see international airports as instruments of some larger economic and social purpose, a situation that leads inevitably to conflicts of interest, with the airports themselves caught sometimes between competing agendas.

The basic approach to these questions will reflect the author's professional background as an academic trained in political economy and management. This means that the approach to the themes and topics in the book will be multi-disciplinary. At the same time, attention will be given to the central issue of transport systems as developmental tools, which leads, in turn, to a consideration of the functional roles airports now play in the macro planning of development policies by national governments.

A further assumption will shape both the themes and the context of discussion in the following chapters. It will suggest that international airports already play a very significant and strategic role within the total development of the world's economies, influenced both by the economic values of their location and by the growing influence of multinational businesses.

Any attempt to cover the airport industry as an international totality would require a multi-volume series. As a matter of personal choice, therefore, it has been decided to address the issues through the study of a specific geographical region, namely

East Asia, which contains a balance of both developed and developing economies. This choice recognizes the growing importance of regional studies, as national states joining free trade areas try to take advantage of their economic geography and bilateral trade agreements, in turn, have become a favoured geopolitical activity of G-8 and Organisation for Economic Co-operation and Development (OECD) member states. This is especially true in aviation, given the growing significance of the region as a major market for aviation services.

In sum, the book will be devoted to the effects of change upon the modern international airport within the East Asian region. It has been partly selected because the author has some working familiarity with most of the states including China. The choice has also been shaped by the fact that the selected location is becoming one of the busiest in the world for air transport services.

I have also been strongly influenced by the fact that my academic and managerial consulting experiences in various East and Southeast Asia countries have actually permitted the practical observation of the growing strategic importance of airports as both national and international hubs. It has also been shaped by the fortuitous consequences of a post-retirement appointment to my University's School of Aviation, which now requires me to take both teach and accept formal responsibility for graduate work in strategic airport development and management.

The book attempts to build on the work of many scholars who are concerned that due cognisance must be taken of the ways in which political, geographical, social, developmental and economic factors move and shape, and in turn influence, strategic managerial and commercial outcomes. It remains to note that the study constitutes an initial exploration of ideas, in the manner of a reconnaissance, of the issues, rather than a final attempt to judge the profound nature of the changes taking place.

Acknowledgements

I should like to thank my colleagues in the Massey University School of Aviation for their often informal inputs into the creativity process; their wide experience has been valuable as a source of information. I should also like to thank my graduate students, especially David Lyons, Mike Haines, Cathy Gilbert. Susan Redmond, Surya Seethepali and Amoy Virkar, whose busy careers in all sectors of the international aviation industry, have often been complicated by the further weight of research assignments. Thanks also to Guy Loft and his team. This is my seventh book, and their kindness, patience and forbearance has convinced me that Ashgate is the publisher to work with.

It remains to thank my wife Beverley whose technical expertise, both in the research phase and in the drafting and preparation of the book, have, as always, been of critical value. With this in mind I would like to dedicate this book to her.

Introduction

The Contextual Focus of the Book

Modern aviation as an international industry possesses many complex and often diverse aspects. It is also the location of very significant problems and crises, as any reading of the contemporary professional literature will attest. Many of these are reflective of an administrative tradition that has been shaped over the years by both national and international laws and regulations. The degree and extent to which such systems should continue to prevail has become a moot point in open skies debates, as increasing market competition presents a declining number of public regulative authorities who are still managing airports, with rising fixed costs, and falling revenues.

In the volatile economic and social environment that is the modern industrial world, the industry's importance as a strategic contributor to the further growth of both mature and developing economies is without question. This book will attempt to both identify and analyse the role of one major industrial institution, the international airport, within a specific context, that of an increasingly multifunctional agency located strategically as both a national and regional player in what is becoming a multi-modal mass transit system serving an increasingly global population. Within the context of further discussion, continuing stress will also be given to the fact that there are many geopolitical issues shaping the contemporary roles of airports as multifunctional agencies serving an increasing range of clients.

This latter issue will inform the opening discussion in the first chapter. The reason is that aviation, like other transport modes, now has to accept that macro issues such as advancing urbanization and the environmental effects of expanding demands for transportation services require an inter-modal and sometimes multi-modal policy response, especially in developing countries.

All the major indicators of the pressures of structural change in aviation have been observable for some time. For it is clear that the influences that are now changing the evolution of the industry are also partially the forces shaping the internationalization process of the industrial and business world. New aircraft technologies have made ultra long-haul transportation viable at a time when there is a consistent and medium- to long-term growth in the demand for services involving much larger numbers of passengers and cargo. In addition, the emergence of the multinational corporation (MNE) is the principal driver of international business, with its wide range of cross border activities taking place in both the developed and developing world. This requires that we view the forces shaping change from the perspective of the economic geographer, as much as from that of the lawyer, the financier or the economist.

The themes to be visited in the following chapters will focus on a specific sector of the aviation industry, namely the international airports that, given their volume of traffic, are identified as major hubs. At the same time, they cannot be observed in isolation from the powerful influences exerted not only by clients, but also by the economic and social changes now reshaping communities, especially in the region under discussion. Attempts at analysis of the primary issues will be influenced by the fact that, as market demand grows on an international scale, the range of services and functions delivered by what have become key hubs in an increasingly complex set of local, regional and transoceanic networks will have also have to face the need to expand operationally in both its own modal and an inter-modal dimension.

This initial imperative is reinforced by the need to recognize that international airports are increasingly services driven and now cover a range of expanding client expectations and resultant new forms of operational activities. They also have a geopolitical dynamic that influences change involving issues that are much larger than the purely internal and technical aspects of their operational environment. This is simply because political decision making, either nationally or supra-nationally is still able to have a profound and regulative influence on all industries, especially when it involves a developing economy, some of which are to be found in the geographical region that is the focus of the book.

There is clearly the need for a wider debate as to the possibility of the presence of larger and more significant pressures to undertake reforms, rather than simply the need to respond to changing conditions of demand and supply. These are being created the primary influences that are shaping the change process, especially in developing countries, where aviation is seen as a tool for the internal development of those communities whose physical location means that they may have no access to markets or to the means for the encouragement of growth in the general prosperity of the national economy.

From yet another perspective, the effects of the globalization of world business on air transport as an international industry remains a matter of considerable conjecture. For the moment it is important to point out that while the positive returns to international trade tend to fluctuate over time, especially for countries whose comparative advantage lies in commodity and not value-added markets, the primary winners today appear to be those MNEs who are able to offer their markets a customer-focused service with the greatest speed and convenience from and to any point on the world's surface.

Some Implications for International Airports in the East Asian Region

The Asia-Pacific rim is perceived to be the future market leader in world aviation with demand expected to rise exponentially during the first quarter of the century. At the core of the development we find China and the various states that constitute the Association of Southeast Asian Nations (ASEAN). This raises some very significant problems in terms of the degree of integration between the individual states in the matter of such contemporary agendas as those invoked by open skies agreements, deregulation, privatization and the ability of airlines and airports alike to access needed capital for investment and market growth.

It is also important to be aware that competition is now emerging between the various international airports of East and Southeast Asia who are beginning to lay claim to becoming regional as well as national hubs. They tend to be led by those countries that have already developed major facilities, or are reaching an advanced stage in their development programmes and are now seeking a positive return on the investment.

The development of the various chapters in the study have been complicated by the high degree of empirical overlap between many of the topics and issues raised for discussion and commentary. The result is that the book is less a sequence of clearly defined arguments and more a series of often interlinked macro and micro issues, which recur in a wide range of the topics. At the same time, themes and arguments will be progressed sequentially through the chapter sequence.

Some Themes to be Addressed in the Chapter Sequence

The Theoretical Parameters of Geopolitical and Economic Change

The book will open with a short discussion on the theoretical parameters that shape both the economic geography and strategic dynamics of structural change with regard to the increasing importance of inter-modal transportation systems, and the strategic importance of the role international airports will play in such systems. The links between both the commercial imperatives and the larger geopolitical issues involving development strategies will also be identified. It will then draw on current reported research through a selected group of authors and will attempt to identify more closely the strategic agendas that airports now face in an increasingly diverse market system. In essence, the chapter will attempt to assess and comment on the fact that major regional airports now have multiple roles.

Particular attention will be paid to the introduction of privatization, as many airports move from public toward private ownership. The initial stages of this trend can be found in the emerging activities of professional agencies operating a given major site under contract either to an official or government authority, or, if the airport is totally privatized to firms or organizations representing the major equity holders.

The Ambiguities of International Airport Privatization as a Strategy

The question of the degree and extent to which privatization should be advanced in the presumed interests of economic efficiency faces a counter-argument in which a continuing role for the public interest is considered to be a viable proposition.

In considering this, attention will be paid to the effects of regulative change through deregulation, ostensibly to permit the creation of a contestable market. The presumption that costless free entry and exit is a market incentive for new entrants will be examined in comparison to the alternative view that incumbent operatives are therefore penalized under such arrangements. This theoretical controversy has a notable importance in the case of airport slots, and will be viewed in that dimension.

The fact that the strategic need to service the airside is being reshaped by the current conflict between, on the one hand, the major carriers, and, on the other, the emergent class of low-cost carriers (LCCs) will also be formally examined. In the United States, there appears to be a deepening competition between the parties, as the major carriers reinforce their holds on regional hubs in order to compete with the LCCs for regional traffic, while at the same time trying to match them through the launch of their own low-cost satellites.

Discussion on this issue will further accommodate the fact that LCC launches are now increasing in number throughout the region. The emergence of LCCs in the East and Southeast Asian region is indicative of the fact that the seeds are being sown for the kind of market competition that the EU law on open skies helped to develop in the 1990s.

The Implications of the Development of Free Trade Agreements in ASEAN and the Significance of Multilateral and Bilateral Aviation Agreements

Further discussion will then raise matters relating to the economic and financial obligations that privatization brings in its wake. The need to be able to satisfy stakeholder obligations through a positive return to the equity holders is a golden rule in business. It now faces those international airports in the West that have been fully privatized. In a real sense, it also faces those countries such as China, where central government has invested massive sums in both new and refurbished locations, but with a different order of expectations based on the strategic requirements of the current National Plan. An attempt will be made to discuss these topics from a geopolitical as well as an economic perspective.

The Emergence of China as a Major Civil Aviation Market

The issues raised in the previous discussion on China's airport policy will be further developed in the context of the role of aviation as a geopolitical tool for national development. In the Chinese case, for example, rapid economic development has been something of a paradox. For while the coastal provinces and those other within logistic reach of the southern ports have materially benefited from economic reform, the peoples of the west and north, some 750 million in number have received only marginal returns from economic reform. Government is well aware of this and sees aviation as an important logistical key for the further development of those regions.

There will be an initial discussion that will focus as a case example upon the strategic issues facing a key region in China, the Pearl River Delta (PRD). Apart from Chek Lap Kok, which is the currently the dominant actor, there are now airports with an international capacity at Shenzhen and Dongguan as well as Macao and Zhuhai. In addition, the new Baiyun International Airport, which has just opened 30 km north of Guangzhou, has a current two-runway capacity, plans for a third and space for two more when needed.

The intention appears to be to create a mega economic zone with an integrated multi-modal transport system. The primary economic purpose for this level of economic concentration is for the system to act as a super-entrepot, servicing the increasingly global markets for Chinese exports. Discussion will attempt to clarify both the possibilities and the problems facing this further example of multifunctional development, using a developmental model based again on the PRD. In a later chapter, the larger question of China's civil aviation policy as a function of its national and international strategies for market reform and liberalization will be dealt with in more historical and administrative detail.

The Complex Dynamics of Urbanization and Peri-urbanization

The identification of the forces now shaping the drive to increase the number of major airports in the region will also be taken up from a dimensional point of view. At the nub of discussion will be the fact that the region that is the central focus of this work is also the site of an urbanization process that is reshaping the nature of cities and their spatial boundaries. The discussion will also consider the fact that major airports are now becoming the centre of active conurbations where the social life of the community overlaps with the economic demands of industrial production.

The fact that the region under study is in the process of being structurally re-shaped is a consequence of the fact that the states of East and Southeast Asia have become a global location for manufacturing production, with China as the epicentre. These developments are now having an important shaping influence upon such essential matters as the location of key hubs, many of which are now being located significant distances from major urban centres in order to service equally major shifts in industrial locations.

The Salient Roles of Logistics and Supply Chain Management

The competitive world of international business now requires that those who would serve the market are aware of the need for speed; as the production system becomes more distributed both geographically and managerially, the supply chain gets shorter, organizational systems for production become more diverse and increasingly driven by advancing electronic technologies. The emerging importance of logistic and supply chain services will be the focus of a later chapter.

The penultimate chapter will attempt a speculative review of future developments ranging widely across a number of concurrent and emerging issues. Some of these signal serious and important problems for the industry. In those categories, such is the current state of the geopolitical and economic tension that any attempt at definitive analysis would be previous in the extreme. The final chapter will attempt a brief consideration of the role of aviation as a key international business and, with it, the essential need for the international airports in the region to become fully integrated into what are emerging as global networks.

Chapter 1

The Changing Role of the International Airport in the Global Economy

Introduction

There can be no doubt that the recognition of the need to make various forms of structural change in the management of the aviation industry is now being increasingly acknowledged on an international scale. This is to a large extent a natural response to the demands being made for new organizational and operational paradigms, research into the causes of change and the need to find ways in which management should respond in the strategic sense. As a result, much attention is currently being focused on specific and internal aspects of the industry. The process is being driven essentially by academics, as well as professional analysts, seeking insights into the possible direction that the necessity for organizational reforms should be diverted.

This study, which takes as its primary emphasis the effects of change on the roles and functions of international airports, obviously fits into that genre, since the sector is facing multiple pressures to adapt and develop a range of new roles and functions. But it will also attempt to identify and comment upon a number of those larger change dynamics that are external to, but underlie and are having a profound influence upon, the managerial and operational parameters of the international airport's roles and functions. In doing so, it will focus on the dynamic changes in the economic geography of the East and Southeast Asian region, and the geopolitical influences that are shaping the aviation strategies of the emergent market economies such as China.

Attention will also be given to the competitive positions being established within the region by those leading edge economies, notably Hong Kong, South Korea, Malaysia, Thailand and Singapore. It is they who are setting the pace in the matter of major hub development. In the case of Singapore and Hong Kong, their undoubted places, way out in front of the other airports, is reflective of the fact that they are already leading members of the global airport fraternity. Before commencing discussion, it remains to note that the term East Asia will be used in a general sense to include Southeast Asia, Northeast Asia and China.

Introducing the Primary Forces Shaping Change

This first chapter will attempt to widen the scope of the discussion on change, its antecedents and its future directions, beyond the purely reactive frame of reference. In doing so, the general focus will be upon the political, structural, commercial and economic forces that are now profoundly re-shaping the world economy and, with it, the international airport, on a truly global scale. The examination of these macro issues will include comment on the historical evolution of those major market forces that have shaped air transport as an international competitive market during the last quarter of the twentieth century. Further discussion will then involve a preliminary examination of a number of topics that will be addressed again in greater detail in the balance of this study.

We will begin by looking at the pressures on aviation being exerted by the increasing needs for inter-modal transport systems as a response to market growth, and the agglomerative tendency for national populations to create urbanization on a massive scale. Attention will then be focused on the major changes that continue to shape international business and trade, and with them the increasing convergence of industries through cross-border location between the various nation states and within specific regions through the creation of free trade areas. This phenomenon has become popularly known as 'globalization' and has actively re-shaped market structures since the early 1970s. The replacement of mixed market economies with free market arrangements has been another important geopolitical shift.

Discussion will then focus on the triadic processes of market liberalization, deregulation and privatization, which are the result of profound changes in the political economy of western developed economies as well as in an increasing number of developing countries, including the former command economies. The impulse to deregulate its market, first experienced by the United States airline industry, will supply the initial benchmark upon which to assess the ways and means that deregulation has seriously impacted on the aviation industry, among many others. Finally, consideration will be given to the ways in which geopolitical imperatives, driven by international agencies or national governments often have the final say in matters of ultimate strategic importance, especially in the areas of policy development and its implementation.

Some Current Imperatives Shaping the Development of Inter-modal Transport Systems

In the twenty-first century, which is already being identified in the popular literature as the age of globalization, international business is increasingly identified through the global branding of goods and services, increasingly competitive pressures from emergent markets, new firms and rising consumer demands. Consequently, the various means whereby passengers and goods are constantly being transferred to most points on the globe are increasingly being identified as the key inter-modal elements of an emergent and international mass transit system. Under such circumstances, the geographical location of any given country or city becomes an important plus or minus in an age where distance from, and speed to, the market have become primary assets in the competitive worlds of international trade and business.

Defining the Region to be Analysed

Given the adage that Hong Kong is at any time 6 hours' flying time from 50 per cent of the world's population, the strategic significance of East and Southeast Asia becomes increasingly obvious. By contrast, Australasia, giving the current state of their aircraft technology, has to accommodate an initial 10- to 12-hour outbound non-stop stage in order to access its growing markets on the Pacific Rim, which is replicated if Europe or the east coast of the United States is the home of the appropriate gateway hub.

The region in question has another important addition to its location due to the fact that is rapidly becoming the home of a significant number of those international airports that can best be described as multi-purpose mega hubs. These are extremely expensive forms of national investment, by countries, including China, whose government perceives aviation as a key tool in its plans for economic development and further growth. Having invested heavily in both domestic and international capacity through the development of new sites and significant airport upgrades, they now have, in keeping with other developing countries, to search for increasing revenue sources and flows if they are to first cover their development costs and then make a profit for the equity holders. They are also motivated by the fact that China is emerging as an economic superpower with an investment strategy in Southeast Asia that both parallels western firms and includes a significant strategy for the development of its aviation industry.

Key sites such as Hong Kong–Chek Lap Kok and the newly opened Baiyun International in Guangzhou are already trying to plan for an expanded role servicing the current and expected growth in international traffic both coming into and passing through the region. It is important to note here that there are some four other international airports in the PRD, which are clearly going to be in direct competition, for inward, through and outward-bound international passenger and goods traffic. All of them without exception are seeking, through the development of their capacity in cargo logistics, to maximize the comparative advantages that accrue from large scale urbanization and the larger industrial and commercial activities, all of which are going on within their geographical proximities.

The spatial parameters of the region that is the location of study have been defined as East and Southeast Asia. They have been chosen while mindful of the fact that cross-border dynamics, both political and economic, tend to blur and make less precise areas defined as having distinct political boundaries. With this in mind, East and Southeast Asia will be used conterminously, throughout the text, with East Asia as the general case. In addition, the term Asia-Pacific will be used where the larger regional context becomes the focus of discussion. This will be very much the case where organizational entities such as ASEAN and Asia-Pacific Economic Cooperation (APEC) are part of a particular discussion.

Aviation's Status as a Service Industry

The industrial classification of aviation as an industry finds it listed in the international market for services. This is a relatively recent categorization, since formal recognition of the scope and size of services generally was only finally obtained when the sector was added as a formal class of business activities to the General Agreement on Tariffs and Trade (GATT) agenda for the Uruguay Round in 1986. As a consequence, it was then incorporated for further advancement in the form of international recognition by member states of the GATT in the General Agreement on Trade in Services (GATS), with the ratification of the Uruguay agreement in 1993. This multinational agreement, which is now administered by the WTO covers all modes of transportation.

There is, however, a vital and important operational distinction to be observed with regard to aviation. The material purpose of the reforms instituted by the GATS, are in fact diametrically opposed to the notion of regulation as reflected in the rules for compliance specified by international conventions such as the Chicago Convention of 1944. Its prime objective is to liberalize the international trade in services and, in doing so, it tends to focus upon the modification through liberalization of the various forms of regulative restrictions for cross-border entry that are usually imposed by national governments. This means that it differs materially from International Civil Aviation Organization (ICAO) and International Air Transport Association (IATA) procedures, with their stress on international conventions covering all member states, which are then endorsed and administered by agencies in those member states. As a result (Findlay et al., 1996), aviation as a service industry has tended to remain largely outside the purview of the GATS administration, with a very limited focus on minor aspects of the industry to be found in an annex to its main conventions.

Within the general definition of transportation services as a multi-modal system, air travel fits the conceptual framework of a modal industry in three specific particulars. First, it creates a close proximity between the producer and the consumer. Second, it is both produced and consumed at the same time. Third, it requires the physical carriage of the consumer from the point of origin to the point of destination.

Air travel is also a highly differentiated product, which can be split into market segments. Movement by air is distinguished by such variables as business schedules, leisure plans and personal matters of significant urgency, cargo requirements and other factors. A clear understanding of all factors relating to the purpose, length and type of a given flight is also important, since the measurement of profit yield is an essential part of the airline's flight planning. Further variables such as short haul–long haul and point-to-point or multiple transfers also shape the mix, while the type of aircraft and quality of lounge accommodation influence business class choices. The development of transportation technologies has increased both pressures and the propensity for transport planners to develop inter-modal systems. From a market perspective, it is also important to note that the growth of the world's largest business sector, international tourism, has identified and created the urgent need for more seamless forms of integration between transport modes.

A further global pressure is derived from a more negative source involving the active relocation of significant human populations. The world is experiencing a rate of growth of urbanization, which is in turn reflected in the growth of cities at something approaching an exponential rate. From an aviation perspective, the pull effects of agglomeration are to be clearly seen (Davies, 2002), for example, in the significant and growing levels of congestion being experienced at major airports in the United States and Europe. According to Davies, cities with populations in excess of 20 million are now well advanced in the process of emergence, and on a truly international scale. The situation is further complicated by the fact that world air traffic is expected to double over the next 20 years and yet again by about 2048.

Some serious problems with capacity management have emerged for the leading international airports, notably in Europe and the United States. These tend to be related to the rate of haphazard growth that has occurred in and around airports, as a result of the increasingly competitive markets for the usage of adjacent land. In fact, the problem has become worldwide in its impact, notably because of the 'knock-on effect' from congestion particularly in Europe. Limited potential for the expansion of existing sites is also becoming a source of considerable anxiety.

In addition, it must be borne in mind that a new airport development, on a greenfield site, is by definition a medium- to long-term project, often delayed in many countries by the need for environmental and resource and planning consents, which ultimately makes the construction process itself extremely expensive. A noted example here is the popular reaction to the location of a new airport for Sydney, a decision on which is becoming increasingly urgent over time. At the time of writing, proposals for a second airport in West Sydney, have met with strong resistance from local authorities, which appears to have left the decision on a site to some extent in limbo.

The actual decision to invest either in new sites or extensions to current locations is a high-risk activity, especially in developing countries. This is because it is often the case that the existence of a centralized multi-modal transport planning strategy may require that any specifications for a new airport be folded into the larger strategic context. From a governmental perspective, it is logical to do so, especially if the plan itself requires the assent of a major international donor seeking the greatest collective benefits from lending for such projects. As a result, arguments in support of a major airport project have to compete internally against what competitors would claim to be equally valid commitments to rail, road and seaport developments.

The 'Mainport' Concept

Suggested responses to what is a growing problem internationally stress the essential complementarity between air travel and high-speed rail services as personified in the French TGV system. This has given rise in Europe to the concept of 'mainports' in which the inter-modal link between air and rail at key terminal points is being developed under the proposed Trans European Transport Network (TETN). The further development of the network of inter-modal airports linked by fast rail services will include Brussels, Amsterdam, Dusseldorf, Munich, Frankfurt, Cologne-Bonn and Milan-Malpesa, coupled with the expectation that the total system will be operative by 2010.

This concept envisages an extended inter-modal solution to the urbanization and congestion problems at international airports, which suggests a linked airport-based terminal system where both point-to-point and transfer traffic can be broken up and serviced by the most efficient mode. High-speed trains, for example, can offer inter-city movements up to a radius of 500 km, although it has been noted that they could take over the business of short point-to-point commuter traffic from local air shuttle services. It has been further suggested (Graham, 1995) that a seamless bi-polar system of rail, high-speed rail, road transport and air links, servicing the national, regional and global services of an international airport, might well offer a more rational approach to the growing demand for air travel, given scarce airport capacity and the need for more new runways.

We have examined, albeit briefly, the effects of urbanization and congestion and the need for a reactive strategy to manage the problems they are creating for air transport. In specific terms, they also exist as a by-product of the growth effects of air transport demand on the physical location of international airports as increasingly multi-functional hubs. The immediate question that now arises is what then is causing urbanization on its international and current scale? This, in turn, becomes a cue to examine what has been called the increasing globalization of business and trade as a primary factor shaping the role of both airlines and airports.

The Significance of Globalization as an Instrument of Structural Change

The traditional definition of a firm engaged in international trade is that of a commercial entity seeking to expand both its production and markets by setting up operations to do so across the national boundary of another country. By the 1970s, the individual firm locating in an overseas market, and with a market growth strategy based on arms' length competition with both local and other international competitors, had given way as a primary agent of international business (Buckley and Casson, 2002, originally published 1977) to the ubiquitous multinational enterprise (MNE).

The MNE was identified in the 1977 study as the primary agent driving post-war economic growth. One of the key drivers propelling MNE development was identified as a structural shift into the production of technologically based goods. This had the further significant flow-on effect since it then raised the general level of investment in research and development. On the demand side the general rise in consumer incomes was coupled to increased sophistication in the use of discretionary incomes. At the same time (Buckley and Casson, p. 102), government investment on defence and 'prestige' projects had significant spin-offs into the defence contractor segment of the commercial sector. These developments were balanced on the supply side by the essential and increased availability of skilled labour.

Today, politicians, bureaucrats, academics and journalists all use a single word, 'globalization' to describe the engine that is presumed to be driving of the international economy. How far the term describes a universal activity affecting all countries remains, however, a matter of some debate The international distribution of foreign direct investment (FDI) clearly indicates serious disparities, with Africa, most notably under-represented in terms of the aggregate level of foreign capital moving into the continent.

For some leading theorists (Kogut and Gittleman, 2002, pp. 435–48) who are supportive of the globalization model, the term has evolved to a stage where it is taken to mean a process of integration or convergence going on at a deep level of social, political and economic life. By contrast, there is a school of thought that argues that the popular concept of globalization as a universal process simply does not fit the reality of what MNEs really do in the conduct of business. They base their counter-argument on the evidence of a geographically uneven distribution of MNE activities across the world of international business. The substance of their dissent rests on the demographics of the MNE as measured by the Fortune 500 surveys. The anti-globalization case argues that the vast bulk of MNEs can be found in a geographical triad incorporating the North America Free Trade Agreement (NAFTA) countries with the United States as core economy, Japan as the Asia leg of the tripod, and the European Union (EU). In a recent development of this theme, Rugman and Brain (2003), after empirical testing using the ratio of foreign-to-total sales as a benchmark, have found that far from dominating markets on a global scale, MNEs largely operate within their home-based markets in each part of the triad.

The range and complexity of these arguments are an indication that no universal and accepted definition of what constitutes globalization currently exists. It also has to be acknowledged that the term itself is fraught with a range of symbolic images, dependent upon the social and ideological perspective people bring to an understanding of what is a very dynamic concept. At the same time, there is a perceptive and growing awareness internationally that convergence both institutionally and culturally is expanding through electronic technology and the growth of transoceanic aviation.

Both the rhetoric and the reality of globalization as the harbinger of what was becoming known in the 1990s as the 'new' economy, came under severe criticism as a decade of consistent growth and declining inflation turned sour. What is now recognized as an international growth bubble had driven the market economy at a rate that made many analysts believe was set for the long term. In East and Southeast Asia, where the western experience of growth was assumed to be replicating itself, key economies, such as South Korea, Thailand, Singapore and Malaysia and Hong

Kong, became, according to the IMF, policy models for less developed countries and regions. Then, in 1997, the region's money markets became subject to a massive correction, as the very large proportion of lending created by private speculation saw the values of currencies destabilized and a serious crisis emerge as speculators sought massive profits by driving down the value of national currencies.

In the West, the requisite hard braking of economic growth as the consequences of market correction brought its own scandals and tribulations with, for example, the virtual collapse of the savings and loan industry in the United States, which destroyed over 1 trillion dollars of pension funds. This was followed by the infamous Enron collapse, which did little to encourage a view of the genuine value and positive effects of globalized free trade, properly managed. These events served to support the views of anti-globalization activists, epitomized in the Seattle meeting of the WTO, which was to symbolize the commencement of what would have been the Seattle Round of further free trade reforms, that global capitalism does not have a genuine moral base.

Despite these various caveats and qualifications, there can be no doubt, as already been noted above, that convergence is still evolving, driven by such macro factors as technology, especially in the fields of electronic commerce and communications, the development of financial networks on a global scale following the abandonment of the Bretton Woods agreement in the 1970s, and the major changes that have taken place in the international division of labour. The key question that has emerged from the perspective of economic geography is how evenly have the benefits of globalization been distributed on a global basis and at what relative speed between various industrial sectors has it developed in different countries?

A useful working distinction is offered at this point in the discussion, which involves a distinction between internationalization on the one hand and globalization on the other. The internationalization of economic activities across national boundaries basically involves (Dicken, 1999) a quantitative process, which expands in the geographical sense the range of economic activities without materially altering the geopolitical status of the countries that are involved. The essential difference that is encapsulated in the concept of globalization, is the notion that the various economic activities taking place across national frontiers progress a stage further when they become part of a functionally integrated international system. At this point, both internationalization and globalization can be said to coexist in time and space (Sideri, 1997) and more importantly in different stages of sectoral and regional development.

The idea that the globalization process is driven by the activities of MNEs, and at different speeds in different markets, serves to endorse the importance of regional integration, especially where there is an operational differentiation between individual labour markets. It allows (Buckley and Ghauri, 2004, p. 83) the fostering of regional goods and services markets to enjoy economies of scale across the states in which they are operating. It has also been argued (Castells, 2001, p. 209) that e-commerce as exemplified by the Internet has an uneven capacity across markets, because it has been built upon an existing system of fibre optic cables. The consequences are seen in the development of server farms and website hotels that offers firms a service to handle the volumes of message traffic that appears to be increasing exponentially.

If MNEs in all of their various industrial manifestations can be considered to be the primary agents that shaped the operational aspects of what we call globalization, we are now faced with another important question. Who then set the needed political terms of reference that went on to shape the conditions under which the MNEs were able to grow and flourish? In order to find out, we must now examine the major shift in their economic policy directions that were adopted by key members of the G-8 group of countries led by the United States and commencing in the late 1970s. In doing so, we have to bear in mind that the pressures for the kinds of changes that will be described below had their origins in the increasing inability of governments wedded to the mixed market economy model to manage their relative decline over time.

The Instruments of Reform: Market Liberalization, Deregulation and Privatization

The links between liberalization as a policy intention and deregulation and privatization as the optional means to strategically re-shape the market process are seamless both ideologically and in terms of government policies. The underlying theoretical presumption, as defined by several of the Nobel Prize winning economists of the Chicago School, supposes that a liberalized or free market is innately more efficient than one that is subject to some form of centralized administration. This essentially classical view of the market gains further support from the assumption that since human beings are driven by the principle of rational maximization in their market expectations, they subsequently order their market conduct in ways that will achieve the maximands that they seek on market entry.

We can put this argument more simply and perhaps with greater force by paraphrasing the nineteenth-century inventor of Utilitarianism, Jeremy Bentham, who is really the intellectual father of the rational maximize. He suggested that when people try to balance scarce resources between a variety of ends, the basis of a market choice, in making their choices they will always seek pleasurable outcomes and try to avoid painful ones.

Given its innate diversity and complexity it is also assumed that the market will respond naturally and to the best advantage of all concerned if it is subject to minimal control. The logic of market freedom is based upon the assumption that the only real source of total information (Hayek, 1960) about the expectations and behaviours of buyers and sellers is the market itself. Since it is impossible for any individual to possess this knowledge as a totality, any random shift in market behaviour is a logical result of the market's response or adjustment to change based on its own total knowledge. It follows that no amount of government intervention will change market outcomes. In fact, such interventions by definition can be very prone to market failure.

While it was theoretical economics that shaped the rational assumptions that inform the concept of the free market, we have to remember at this point that the actions that created the processes of liberalization deregulation and privatization that has since swept the post-1970s world had their origins in the geopolitics of the major western states. The bureaucrats who managed the evolution toward the free market economy of today were in effect directed by the leaders of governments for whom the theoretical underpinnings on neo-liberalism were a useful form of justification, when criticized by their parliamentary oppositions.

The Political Economy of the Free Market

In order to fully understand the political events and strategies that shaped the re-emergence of market liberalization, deregulation and privatization, it is necessary to first place them in their historical context. The period from the beginning of the 1950s until the mid-1970s has been characterized by some economic historians as 'the great boom'. It was a time in which capitalism was 'managed' by macroeconomic strategies that placed central government in the role of the key player. Britain, Germany and France all developed their own national forms of planning strategy, while the United States engaged in a policy of regulation. The result was an extended period of economic growth, which was to come to an end both symbolically and literally with the Organization of Petroleum Exporting Countries (OPEC) oil crisis of 1974.

From a theoretical perspective, one of the major intellectual forces that shaped the post-war period was John Maynard (Lord) Keynes, whose work both as an economist and as one of the architects of the post-1945 drive to reconstruct the international economy dominated government thinking throughout the period. Keynes was of the generation who had taken an active part in the discussions post-1918 at Versailles. He became a very strong critic of the Treaty of Versailles, which historians have since taught us laid the foundations for the 1939–45 conflict. In addition, he was one of the seminal influences on the Bretton Woods agreement of 1944, which created the framework for interventionist policies by central government. The impact of what became known as Keynesian macroeconomics resulted in the commitment of the western economies, post-1950 to a managed and welfare orientated form of capitalism.

The period of the 1970s was to see the long drawn out collapse of the modern mixed economy, as it began to be plagued by serious problems. Unemployment rose and so did international inflation, while productivity fell significantly. The end of the 1970s was then to see ultimate political power in the United States, Britain and Germany take a 180-degree shift. Symbolically, the elections of Ronald Reagan, Margaret Thatcher and Helmut Kohl, signalled a political shift in ideological leadership from centre left to the centre right.

What followed from a theoretical perspective (Canterbury, 2001) was a counter-revolution, which was marked by the return of classical economic market theory, but in a twentieth-century guise. While the concepts for which Keynes became famous are irrevocably associated with his alma mater, the University of Cambridge, the intellectual forcing ground, as we have already noted above, for the theory that underpins modern neo-classical economics is the University of Chicago.

The principal axiom of the neo- or new classical economics that began to influence public policy in the West was the essential ineffectiveness (Hayek, 1960) of policy interventions by the state. This meant in operational terms the market in its primary role as the instrument of exchange between buyers and sellers was far too large, complex and subject to random changes to be effectively managed by governments. Such views were to influence the policy makers and those charged with the technical design and implementation of the political decision makers most profoundly from the end of the 1970s onward.

The policy instruments chosen by the 'New Right', a phrase later coined to identify the increasing number of reform minded and centre-right governments that were to come to power in the 1980s, called for the liberalization of much of the administrative activity that had for so long been the province of public sector management. Market reform was to mean the active and extensive withdrawal of the state in terms of the design and implementation policy directions. But it also had a secondary effect brought about through the systematic deregulation of what had become the state sector. Many functions and activities that the mixed market had required the public service, both national and local, to manage became subject to tenders for services, often by private sector firms and agencies.

But the change process did not stop there. The logical next step in the 'freeing up' of the market, saw former state-owned industries become subject through the process of market privatization to private ownership. It is a quirk of history that transportation and aviation in particular were among the first industries to be privatized, as the following examples will attest. A classic example of the evolution of stages from public ownership, through deregulation and onward to privatization is found (Botton and McManus, 1999) in the case of the deregulation and later privatization of the British Airport Authority (BAA).

In a very real sense the change process was to become active prior to the Thatcher era, which began in 1979. The birth of the BAA occurred when attempts to manage airports the size of London's Heathrow, through a nationalized central bureaucracy reporting to a government minister, went out of official favour. The result (Botton and McManus, 1999, p. 274) was the Airports Authority Bill of 1965. In 1987, after a period of major strategic acquisitions, BAA became a plc (public limited company), and was launched on the London Stock Exchange as one of the first acts of privatization by the British government of the day. In a way, the move was made easy by the fact that, in its previous incarnations, the BAA had enjoyed a high degree of autonomy from ministerial interventions.

In the US case, the airline sector became the focus of a deregulative strategy that has its origins in the early 1970s. The prevailing political mood was shaped, post-Watergate, by the Ford and then the Carter administrations, which began the search for reform of inefficient administrative practices in government.

A set of fortuitous circumstances created by the need for a Senate Committee to undertake a survey of administrative and regulative processes in government identified the Civil Aeronautics Board (CAB), which had been created in 1938, as the first organization to be targeted for review. The CAB had its origins in an attempt by the Federal government to resolve the chaotic situation facing the struggling airline industry as war became more and more probable.

By the mid-1970s, the CAB had become little more than a cabal run on behalf of government in which the federal regulator as a political appointee and the airlines enjoyed a very comfortable relationship. Routes were awarded to the various operators on what was called the 'Dogs and Plums' system. An airline would be required to service a loss-making Dog-route, and would in return receive a Plum or money-making route as compensation. The loser throughout was the passenger, since airlines simply offset the losses made on thin routes by higher prices. It was a strategy that had shaped the prevalent view of costs by US airlines up to the time of the CAB review.

In 1997, President Carter appointed Professor Alfred Kahn, who was a leading authority in the economics of regulation, to the chair of the CAB and he soon made it clear that he was not happy with the state of affairs at the agency. It was Kahn who once defined an airliner as 'a marginal cost with wings'. His attitude as a devout marginal theorist is reflected in the fact that he began the change process with a determined assault on the pricing system through the introduction of fare discounting. The impact of this decision was to see over 50 per cent of economy class passengers paying what became known as 'peanut' fares.

Despite inevitable resistance from the airlines, deregulation was soon a fait accompli. In October of 1978, the Airline Deregulation Act was passed and the mainstream carriers were free to set their own fares in direct competition with each other. They were also free to either enter or exit a given market at will. In a symbolic sense, the era of rigid control, complex bureaucracy and route trading finally came to an end when the CAB was closed down in 1985 and the responsibility for air safety was passed to the Federal Aviation Authority.

The Post-deregulation Realities

The strategy introduced by A.E. Kahn as Chair of the CAB was strongly influenced by what had emerged in the economic literature (Baumol et al., 1987) as the theory of contestable markets. The concept was based on the assumption that deregulation (Stiglitz, 2003, p. 101) would allow for the possibility that the entry of fresh competition into a market from new airlines would, in turn, exert a strong discipline on the current operators not to raise prices. The model went on to assume that even where a single airline flew a given route as a quasi-monopolist, the potential fear of zero profits and low prices, from the appearance of new entrants would deter the said airline from charging a monopoly price. Unfortunately, in the event, the reality of the new market simply did not match up with the theoretical presumptions being advanced.

It is true that a number of new airlines, such as People Express and New York Air, did enter the newly deregulated domestic US market, but of these early new entrants, only Southwest survived and prospered. In fact, as competition increased between what became fewer but much larger carriers, a number of well-known names such as Eastern, Braniff and Pan Am followed them out of the market. The survivors went on to ride the wave of increasing international demand for air services, as prices fell, by expanding into the transoceanic market.

Competitive market theory was simply not ready for the development of the hub-and-spoke system, coupled with extensive slot control at important hubs by the major airlines and the ability to drive out new competitors, who tended to be undercapitalized at the time of entry. The instruments used to do this was the highly controversial predatory pricing and predatory scheduling. These techniques will be examined in more detail in Chapter 2. Professor Kahn did admit over a decade later that, while the introduction of competition had lowered prices and made air travel accessible to all in America, there was still the question (Kahn, 1988, 1993) of whether or not the reforms had gone far enough. It is interesting to speculate at this point that he might have had in mind the relationship between a deregulated airline market at the domestic level and the fact that the international market still imposed regulative requirements, soon to be the focus of a growing 'open skies' debate.

Deregulation may be defined in terms of the transfer of control and management as a partial response to the political drive for more market freedom. This is because the practice of many governments, in deregulating state-owned enterprises (SOEs), often stopped short of the disbursement of total control over a SOE's management and its equity. The strategy employed would often introduce the notion of a 'golden share', in which a national government ensured that by holding a significant proportion of the equity, there would remain both a modicum of final control and a revenue stream accruing as an annual return. In addition, as demonstrated above in the case of the BAA, while management would enjoy a degree of autonomy, the need to report to government would be retained.

By contrast, privatization as a general principle, involves the sale of all the equity, and with it any controlling interest to parties wishing to either expand their current operations or to enter the industry for the first time. As a consequence of the transfer of the total equity and control, often to the highest bidder and through a brokerage arrangement, the SOE, literally becomes legally and operationally a private firm or organization active in the private sector. As a result, privatization creates the ultimate form of market freedom.

There has been growing pressure for some time (Doganis, 2001, 2002) for airline markets to be liberalized, as witnessed by the general arguments for and against 'open skies agreements'. While progress has been made toward more liberal policies of deregulation since 1978, the process of further liberalization is complicated by the fact that a given country's sovereignty over its airspace can be offset in a very real sense by the geopolitical nature of regulatory controls that are vested in international agencies such as ICAO and IATA. The rest of this chapter will examine some of the implications of this situation from a geopolitical perspective, prior to further and more detailed discussion in Chapter 2.

The Geopolitical Issues Facing Market Reform in the Aviation Industry

The formal definition of geopolitics suggests that its primary concern is the connection between geography and the international affairs of a national state. In a very real sense, the history of aviation as an industry has been shaped in the managerial and administrative meaning of the term through the formalization of general principles by international agencies such as the ICAO. Regulative procedures after their formal promulgation are then traditionally followed by further interpretation at the state level, undertaken and then administered by national civil aviation agencies set up for the purpose.

At the core of those regulations that apply to the international movements of aircraft on an increasingly global scale may be found the regulative concept, first endorsed by the Chicago Convention of 1944, that a specific nation state, as defined by its geographical boundaries, has sovereignty over the air space that covers those boundaries. Within the parameters of this ruling, it has become common practice for countries to enter into bilateral agreements with regard to the various freedoms that define the operational extent to which a given airline can enter and use the facilities of another state on a reciprocity basis.

The most advanced developments in open skies agreements thus far can be found in the EU, which, in 1997, completed the third and final legal step in the opening up of free movement by airlines between those member states that form the political union. The expansion of traffic that followed saw the emergence of a plethora of LCCs, which, it is no exaggeration to say, has altered the entire nature of the airline industry in Europe. Using the seminal model created by Southwest Airlines in the United States, the no frills, low-cost operations that have emerged, such as easyJet and Ryanair, have dominated what is best described as a new form of low cost mass market for air travel.

Today, the growing market weakness facing major carriers adds to the considerable and growing pressures for the further liberalization of the airline industry. The ultimate expectation of greater market freedom is symbolized by what has been called the 'eighth freedom'. This, in effect, would permit major airlines not only to enter into market competition with national flag carriers and within another country, but also to follow common MNE practice through the purchase of interests in that country's airlines and airports. There is evidence as we shall see in the next and subsequent chapters that progress is being made towards greater liberalization. But there are some very clear indications that there remains a very considerable problem with regard to the perception of some of the major governments as to how liberalization should be achieved.

For example, the United States has persistently refused to abandon its traditional use of bilateral arrangements as the primary vehicle for expanding its strategic links internationally. This led during 2004 to a crisis in which the European Commission publicly chastised America on the grounds that its persistent use of bilateral agreements with EU member states breached the Single Market Act, a core piece of European legislation, an issue which will be taken up in more detail in a later chapter. A major consequence of this announcement has been the breakdown of discussions that were taking place on the possibility of a 'Trans-Atlantic Open Skies' development between Europe and America.

The event demonstrates another important aspect of geopolitics: the fact that decisions relating to the liberalization of the airline industry are often subservient to a larger political objective. The European Commission's proactive decision to liberalize airlines forms a logical extension of plan to develop a single union of states joined by common links such as currency and, eventually, political status. In this, a single economic market in which all industries freely interact and compete fits within the set of larger geopolitical objectives set by Brussels. This has the effect of making any development of an open skies agreement subservient in the overall strategic sense to the building of a single political entity, which currently comprises 25 sovereign states.

By contrast, the United States has long refused to permit airlines to exceed the bilateral limitations on their activities in American airspace, particularly in the matter of carrying passengers from gateway hubs through domestic airports, prior to exiting outbound to their airport of origin. Further, the severe constraints on the ability of foreign interests to take an equity position in an American airline is reflected in the fact that such parties are permanently restricted to a maximum shareholding of 49 per cent, but with a limitation of 25 per cent of the voting equity on any airline board of directors. Finally, any attempt by a major foreign airline to merge with an American counterpart can become immediately subject to possible federal legal action, under the Sherman Anti-trust Act of 1890, on the question of potential monopoly status, which is a breach of that anti-trust law.

The Impact of Deregulation on International Airport Development

From an international airport perspective, there are a number of operational advantages that emerge from their increasing liberalization that are not shared with airlines. The ability to plan for an expanded range of facilities and revenue-earning opportunities is reflected in the increasingly multi-purpose activities of the landside. These range from conventional service-related amenities to the decision by the South Korean government to create a free trade zone adjacent to the Incheon 'winged city' site.

At the same time, a degree of symbiosis is being developed for another important reason. The forces that increasingly shape the strategic location of business are increasingly supply chain driven. This is because the primacy issue of speed in response to the customers' demands has become a major strategic advantage in increasingly competitive markets. There is also, as a consequence, the growing tendency for firms to geographically locate (Porter, 2000) as forms of industrial 'clusters' so that they can maximize the benefits of network interaction and the possible outsourcing of production of goods and services. In doing so, they need to be able to access short and value-added supply and service chains in order to move the output of products internationally and at speed. In addition, as firms outsource, they tend to increase the range of possibilities and opportunities to expand the range of their activities.

The success of BAA offers an example, since it has become a multi-functional business, with an income from the international management of aviation real estate that now comfortably exceeds its airside revenues. This means that in direct competition with firms specializing in professional airport management, it manages under contract airports in Australia and the United States. We will return to further examples of this type of outsourcing in a later chapter.

The notion of clusters, as operational networks of businesses, each contributing an incremental value to the sum of a specific product or range of products, finds its origins in medieval Europe and most notably in Italy. The development of the modern cluster model by Porter finds its first expression in earlier studies (Porter, 1990) in which he developed his famous diamond model of market competition between firms. This has been adapted by numerous aviation industry analysts as a model comprising five major forces that confront the airline industry today. They have a practical use at this juncture, since they allow discussion to be summarized on a macro issue basis. They also reflect an element of urgency with regard to the need for airports, as well as airlines, to respond strategically to the range of changes now re-shaping the aviation industry.

1. Increasing globalization and adaptation to the processes of convergence;
2. Increasing regionalization brought about by the increasing stress on location as a comparative advantage;
3. Economic incentives for consolidation in order to obtain economies of scale through alliances and networking;
4. The relative speed of market liberalization, which is now driving some of the important economies in the region under review;
5. The need to remove current limitations on market competition, in order to allow airlines and airports to take full advantage of the increasing internationalization of economic activities such as tourism.

Figure 1.1   Porter's five forces theory applied to the aviation industry
Note: For a major overview, see OECD Secretariat (1997).

Some of the forces of change now find expression, on the airside in particular, in the declining economic state of the traditional full-cost airline. Complexity is further compounded by the emergence of LCCs, who now strongly challenge the legacy carriers in their own domestic markets. The emergence and growth of strategic in-market alliances between many of the major players in the international sector is also an important influence, seen by some commentators as a portent for a future which may, in their opinion, be dominated by mega-carrier airlines.

Chapter 2

Structural Changes in International Business: Implications for the Airport Industry

Introduction

Over the last 20 years, the structural changes that are reshaping modern business have tended to create new directions for research into the nature and operational dynamics of firms and organizations. The profound influence of information technologies as key change agents is easily seen in the enormous technical strides made by what are now global communication systems. Their advancement as a universal means of communication has been materially assisted by the fact that network expansion has been accompanied by very significant falls in the both the cost of services and the final price of system usage to the consumer.

The rapid evolution of electronic technologies has also signalled the emergence of a global market for knowledge as a tradable good in its own right. The advent of what is popularly known as e-commerce has had a somewhat chequered career, as the rise and fall of the dot.coms will attest. But it remains a very powerful if now a less strident element in the continuing development of convergence in the international economy.

One of the most important developments in the various studies of the influences now being exerted on organizational structures as a result of changing patterns of market demand has been the move away from the conventional emphasis on the vertical integration of managerial control, with its various gradations of authority based on power and status. For some time now, the international trend has been toward flatter and more horizontal systems of management.

There has also been an increasing recognition of the value of some degree of autonomy and control of work at the production or customer interface. This flows directly from the growing awareness that business organizations have their own identifiable 'cultures'. The word culture implies that businesses are social as well as economic entities and produce their own forms of cohesiveness, loyalty and solidarity. Such collective values are a source of very important strength, since they create in individuals the incentive to work for the common good.

Both geography and demography have played a major role in this development. The location of research and development, which is essential for increasingly high-technology products, tends to require some degree of proximity to the production of intermediate goods and services, such as venture capital, as well as to some final assembly sites. Since many of the intermediate factors are outsourced, shorter supply chains are required for efficient production. This has led to the tendency for various contributing firms to locate adjacent to each other on both a local and regional scale. It has also led to the identification of specialist roles and functions by dedicated service firms, who then supply their unique skills under contract and across the mainstream producers group.

The prototypical example of this development is of course, Silicon Valley, which has now been replicated both nationally in the United States and internationally. The process of deepening convergence can also be found in what are now called 'clusters' (Porter, 1998), a word that has entered the business lexicon to describe the increasingly integrated business networks that have developed as a result of the emergence of new technologies.

The traditional role of manufacturing has also been undergoing major changes, especially in the West and under the expansive pressures of technological progress. As a consequence, the traditional large-scale linear form of production, with its primary stress on the need to reduce the unit cost of labour, has given way to the active recognition of what is called (Brooking, 1997) intellectual capital. From a purely business perspective, this is now being recognized as the key dynamic underpinning a firm's culture and it comprises the net value of the collective skills and experiences of the labour force, which has accumulated over time. It is important to also recognize that such capital has been found to contribute a significant proportion of the total revenue streams of the firm.

The concept of intellectual capital has also been linked with a growing emphasis on cooperative arrangements at work, such as team-based task assignments and other forms of activity that stress the importance of individual autonomy in decision making. The intention is to give strategic emphasis to the total contribution that is made to the final outcome by all members of the organization. The notion of the company as a dynamic network of linked and common interests has also taken hold in many firms. Increasing competition has also acted as a spur toward another major shift in strategic focus. This expands the concept of the stakeholder to include the customer as the final arbiter of quality of the firm's performance. As a consequence, many firms now identify the client or customer as the core focus of the business.

Implications for the Aviation Industry

The preceding discussion is intended as a simplified description of the kind of market relationships that are now common in western business. It immediately raises the question as to how far the aviation industry will have to change its traditional strategies in order to obtain the kinds of competitive efficiencies and increased income streams that it is actively, and in the case of the legacy airlines, desperately, seeking.

In order to do so, there is a need to examine some of the major influences that have shaped both the airline and the airport sectors on an international scale since the passing of the United States Airline Deregulation Act in the pivotal year of 1978. In the case of the airline industry, the semantics of discussion will involve the current practice in the literature of using terms such as mainstream, network, major and legacy carriers to distinguish traditional full-cost operators from their low-cost and independent competitors in the point-to-point and city-pair sectors of the market.

Airlines and Airports as a Mutually Dependent Network

When we examine the basic parameters of both the airline and the airport industries, we know that they are best described in the operational sense, as a network of nodes joined by an often complex system of linking routes. On this empirical assumption, it is also possible to describe them as co-existing in a mutually dependent relationship. The network concept can also be identified as a basic paradigm (Fridstrom et al., 2003), since airline routes intersect with airports as nodal locations. From the perspective of both demand and supply, such networks offer large externalities in the sense that the costs and revenues as they exist between different routes on a point-to-point service become interdependent through network scheduling. These conditions produce large economies of scale, scope and density.

Projections of both passenger and cargo growth signal, despite the essentially cyclical nature of demand, a consistent expansion on an international scale over the short to medium term. If the strong evidence of increasing technological sophistication in aircraft design and manufacture is added to passenger growth potential, there is literally no reason why (Calder, 2002) future global networks will not emerge as the number of locations capable of being reached through ultra long-haul services are increased. The ability to be at some point on the earth's surface approximately 24 hours after leaving a designated airport is technologically feasible. A more complex question, however, still remains unanswered. Can the development of both existing and new airports of the required functional magnitude and operational complexity keep pace with the increasing number of airline services that might wish to use them? The examination, in later chapters, of the developments taking place in East and Southeast Asia will attempt to offer several examples from the region as a potential model.

During the balance of this chapter, primary attention will be given to a selected number of strategic issues, some often contentious, that have shaped the aviation industry, and the role of the airline in particular, following the post-1978 decision to deregulate inter-state air passenger transport. The operative shift to hub-and-spoke services will be considered not only for its positive effects. The failure in practice of contestable price assumptions to produce the needed market-based regulation of competition in an equitable manner will also be considered.

In addition, the vexed and ambiguous question of predatory pricing and other attempts to regulate hub control by the major carriers will be examined. The balance of the chapter will focus on the emergence of the LCC as an increasingly international phenomenon. Finally, the increasing pressures for 'open skies' agreements and further market deregulation, will be examined against the assumed demise of the 'legacy' carriers through potential bankruptcy, and will serve as a prelude to Chapter 3.

The Ubiquitous Role of the Hub-and-Spoke System

The rapid development of networks was, of course, a direct result of the emergence of what was intended to be a competitive market, after airline deregulation had become an operational reality. Its most enduring model remains the hub-and-spoke system, a radical departure from the traditional point-to-point non-stop systems that existed prior to the introduction of market deregulation. The practical efficiencies of hub-and-spoke systems, have been well described by a very large professional literature and do not need rehearsing here, except to say that they range across a variety of cost-saving and revenue-generating advantages. For example, the provision of strategically located spokes may also be extended to cover those thin routes that in an earlier and more regulated environment did not really justify a regular service.

Suffice it to say, the rapid international expansion of the deregulation process outside the United States was accompanied by various manifestations of the hub-and-spoke system, particularly as the cost of flying began to decline and the demand for air services grew. It had been a primary requirement of the United States Congress that the extent and scope of deregulation should become manifest in an active market for services driven by open competition. In other words, the new market should be free and contestable. It is important to observe at this juncture that this was a period of theoretical justification in economic thought, which strove to give liberalization a strong intellectual as well as practical justification. It was coincidental no doubt, but timely, that the period immediately following the deregulation of the airlines was to see the emergence of an important economic model, contestable price theory, that fitted the reformist assumptions for the post-1978 development of the aviation market.

Diminished Expectations of a Contestable Market open to Competition.

The theory of contestable markets, reference to which has already been made in Chapter 1, takes as its first premise the notion of competition as an implied threat, rather than an actual reality. While the abstract nature of the model is conceded, it is suggested that a contestable market would be viable (Donne, 1995) providing that three conditions could be satisfied.

First there must be no disadvantage accruing to a new entrant vis-à-vis existing firms. This assumes that new entrants have access to the same technology as well as perfect information relating to input prices, products and market demand. The second condition would require that there be zero sunk costs, which means that all costs associated with the physical requirements of entry are fully recoverable. It also means that an exit from the market by that same firm would be costless, less any amount caused by depreciation during the period of the firm's location in the market. Finally, there is a requirement to recognize that the post-entry lag (which is the time between when a firm's entry is known by existing firms and when the new entrant is able to supply the market) needs to be less than the price adjustment lag for existing firms (the time between its desire to make price changes and its actual ability to do so).

These assumptions attempt to respond to the situation where (Hanlon, 1999) a deregulated airline market is susceptible to 'hit and run' or minimal sunk cost and short-run activities by new entrants. The presumption follows that even if there is monopolistic or oligopolistic competition active on a given route, there will be a disincentive to raise prices because of the implicit and strong threat of competition from new candidates for market entry.

In the event, its discussion has proven highly contentious, with the supporters of the case for contestable markets arguing that there is a degree of theoretical fit. They base the claim (Graham, 1995, p. 71) on the fact that an airline's sunk costs are relatively low in relation to their fixed costs. Unfortunately, this theoretical assumption is offset by empirical evidence. Fixed costs, for example, are relatively small and often, using airports as an example, covered by either government or the privatized airport company. On the other hand, while the cost of fleet acquisition is high, it is in fact a variable asset if purchased and subject to definition as a fixed cost only if it is wet or dry leased. Contestable market theory lost a significant proportion of its explanatory value after deregulation in 1978. This is reflected in the fact that a significant number of new entrants were forced to leave the market through bankruptcy. This fate was shared by a number of legacy carriers who simply could not adjust strategically to a competitive market.

Recent empirical work in Europe (Pitelis and Schnell, 2002) on the perceptions of airline managers with regard to mobility barriers post-market deregulation also includes tests of the contestability hypothesis. The findings, which support earlier American research, confirmed the existence of endogenous mobility barriers, and the belief amongst managers that they do have important strategic implications.

Since such assumptions undoubtedly shape managerial decision making, the question can now be asked, in what sense are these barriers endogenous and how do they work in practice? In response to the development of the hub-and-spoke system, major carriers, led by the United States, took an overall strategic view of the location of their various hubs.

The result of this approach saw both domestic and international airlines using major nodal sites on their route networks, in order to establish a dominant operational presence. With the consistent expansion of this practice over time, we now find most of the big international hubs specifically associated with an airline that is also the national flag carrier for the country. Within the East and Southeast Asian region, the pattern shown in Figure 2.1 emerges.

- Singapore Changi International Singapore International Airlines
- Bangkok Don Muang International Thai International
- Kuala Lumpur King Abdul Aziz Malaysia Airlines
- Manila, Corazon Aquino Philippines
- Jakarta International Garuda International
- Hong Kong Cathay-Pacific
- Beijing Capital Air China

Figure 2.1   National flag carriers located at the major East Asian airports
Source: Hufbauer and Findlay, 1996.

The presence of a major carrier as the main user of allocative slots at a given location raises some important issues given the increasing pressures on airport capacity already raised in part in Chapter 1. International practice with the noted exception of the United States, tends (Graham, 2003, p. 121) to be based on industry self-regulation, using the IATA Schedule Coordination Conferences, which are convened on a twice annual basis involving some 260 airports.

The system operates at two levels with the first schedule facilitation process reliant on voluntary collaboration between user-carriers, as capacity utilization approaches its zenith under existing physical limitations. The second process of full coordination is balanced where demand exceeds capacity by procedures for slot allocation. Results are obtained by:

- *Grandfather clauses*, which allow regular users to book forward, providing they operate at least 80 per cent of their scheduled flights. The system is based on the principle of 'use it or lose it'.
- *Switching* between domestic and international schedules is also permitted, while preference between users is also applied, where one airline operates daily as opposed to less frequent services.

An increasing number of large airports are now appointing dedicated coordinators, usually the national carrier of the country concerned, to take total oversight of all slot allocations. As a consequence, this key activity is becoming increasingly subject to a body of increasingly standardized practice. The trend toward increasing formalization has led ACI-Europe to develop a model for no less than six forms of 'contractual relationships. They involve the following set of formal arrangements:

- *basic agreements* that elaborate the services to delivered in return for the financial charges incurred by carriers;
- any additions to that basic agreement that might be negotiated;
- *facility agreements* where specific locations and services on the airport site are used;

- *unilateral commitments* to achieve defined operational standards by the airport management;
- *bi-lateral commitments* by operators and users to reach defined quality of services;
- *strategic partnerships* (SPAs), which cover a range of developmental as well as functional activities.

All of these shifts are reflective of the increasing pressures for market-driven strategies such as privatization to be used as a strategic response to the increasing complexities facing the aviation industry at large and airports in particular as service agencies. At the same time, there is a distinct problem of imbalance that arises at this point. The United States, as the largest actor in both domestic and international aviation, remains outside the general trend. The formalization of such changes remains constrained by the long-term existence of anti-trust legislation.

Despite being host to some of the largest firms in the world, the United States has always treated corporate monopolies as essentially hostile to the notion of market freedom. This means that the kinds of agreements already specified would be challenged as collusive activities leading to monopolistic competition in all markets.

The Controversy over Market Manipulation and Predatory Pricing

The concept of predatory behaviour has applications in both economics and the law. From a strategic perspective, it involves a willingness (Holloway, 2003) to limit profits in the short run, on the assumption that market power will be achieved and, with it, the ability to make greater gains in the long run.

Applied to the airline industry, it involves the willingness of a carrier with a large and integrated network to set prices at either below costs or at a level substantially below what the market will normally bear. The conventional example used to demonstrate the strategic impact of predatory pricing usually takes the advent of a new entrant into a specific market as the causal factor. The strategic response by the market incumbents, either singly by the leader or in concert, is to undercut any price advantage enjoyed by the new competitor until such a time as the new and presumably under-capitalized venture is abandoned. At that point, the established market leader again, either as the dominant party or in collusion with the regular competition, is free to return prices to some normal rate of return, often with a discrete upward adjustment over time.

The Case of Air New Zealand versus Kiwi Air

A notable example of such manipulative activity can be found in New Zealand with the launch of a small private carrier called Kiwi Air. This pioneer low-cost airline, using a single aircraft, began to offer flights in direct challenge to Air New Zealand, which covered both domestic and international services. As a consequence Freedom Air was launched to compete with Kiwi and the undercapitalized new operator left the market. With some irony, Freedom Air has continued to service a growing trans-Tasman market.

Other Forms of Predatory Manipulation

Price Signalling

Price signalling is seen as a form of price fixing, where the intention of the parties is to stabilize prices by common consent. The legal onus of proof, however, requires that a distinction be recognized between the exchange of price information on the one hand and the intention of the parties on the other. According to one authority (Morrison, 2004), the problems of definition in predatory pricing have always been rendered difficult by the traditionally static nature of market analysis. These are further exemplified in the airline market by a consistent and dynamic evolution over time, especially since 1978. The fact that both full-cost and low-cost carriers now co-exist and compete in the same market tends to complicate the traditional assumption of the predatory pricing model, that new entrants are essentially the same kinds of firms as the incumbent community in the market.

Allocation of Customers, Markets or Territories

This practice tends to be a sub-set of price fixing in the sense of mutual agreements amongst what are technically independent competitors. These activities are considered to be unlawful per se, but in the case of airlines, operational reality complicates the issue. In the case of city-pair markets, the vast majority are dominated by a single airline and can be defined as either monopolies or duopolies.

Controls over Route Scope and Density

The economies of scope and density have clearly shaped the nature of post-deregulation competition. Given the existence of fewer but larger carriers there is a natural tendency to see the market (Fridstrom et al., 2004), as one that is subject to high levels of control by the big players. In Europe, they estimate that in 2000, almost 75 per cent of all non-stop routes were either a monopoly or a duopoly operation. Recalling that the network system in Europe predates events in the United States, they also note the further complication that any assumption that changes has been an autonomous market process that has to be modified. The regulatory framework in Europe has been driven by the geopolitics of a larger community development for member's states of the EU. In addition, national flag carriers tend to occupy a commanding position both bilaterally, in terms of inter-state arrangements, and as noted above in the management of slot allocations.

In the US case, it has been argued (Brock 2000) that the industry has experienced three major developments of predation in the following forms:

- increasing concentration through domestic and international alliances;
- predatory pricing by network airlines aimed at the growing competition from low-cost independents;
- collusive pricing by network members through mutual agreement to lock out new entrants.

A countervailing argument against this form of criticism has been mounted (Lee, 2003) in which tests using US Department of Transport data were used in an attempt to validate these assumptions. Lee's conclusions run strongly counter to the Brock argument. He found that the national share of the network carriers has in fact declined and that, overall, prices have fallen steadily since 1990. The conclusions also found that both alliance activity and predatory pricing appeared to have little foundation and were lacking support in the literature.

While such academic debates encourage the search for a deeper understanding of the competitive forces that are reshaping the airline industry in America, they continue in a situation of manifest uncertainty (Levine, 2003) with regard to the future of the legacy networks. The fundamental weakness of the mainstream carriers is reflected in their apparent inability to develop a positive strategy to escape astronomical market losses over increasing time. Their attempts to use Chapter 11 bankruptcy rules as a strategic tool, notably in the area of the withholding of contributory payments to staff pension funds, has resulted, some legal critics would argue, in what appears to be an emerging threat to the much larger concept of contractual employee rights to post-career pension.

The ability to obtain potential investment from the general money market, presuming that the current regulative environment is liberalized to allow for such a contingency, is also problematic for the legacy operators. The current market capitalizations of airlines finds that the leading low-cost independents, have a much healthier reputation. Southwest for example has a market capitalization at US$10billion, which is larger than American, Delta, Northwest and British Airways combined. In a very real sense, low-cost carriers do have a degree of advantage over the legacy group, because they come to the market unencumbered by a long history of failure to make a return on the equity invested. By contrast, while undercapitalized new entrants have gone to the wall, post-1978, Southwest has become a management school textbook example of progressive and innovative management.

Low-cost Carriers: Are they the Airlines of the Future?

There can be no doubt that the most profound influence upon the changing fortunes of the network carriers, has been the advent of the low cost independent airlines, many of which have deliberately, like Ryanair, modelled themselves upon the template first shaped by the now legendary Southwest, which has, in October 2004, signalled 54 consecutive quarters of profit growth. There can be no doubt that the emergence of low-cost, no-frills airlines has radically reshaped the industry. Led by entrepreneurial managers, with something of a gift for media exposure, as individual players, their annual growth rates are often as high as 25 percent. From an airport perspective, their arrival in the short-haul point-to-point market has proven to be, to some degree, a mixed blessing. This judgement applies to both newly commercialized and privatized airports that are prepared to offer large discounts and to lesser used and often municipal locations. The seminal question for airports entering into such arrangements, however, lies in the key area of risk.

How long, for example, will the agreement last, especially as in the United States the partner carrier becomes bankrupt or simply withdraws from the route? This is especially true where the airport is reliant upon a single airline. The presence of legacy carriers introduces risk of revenue decline, given the essentially cyclical nature of that market. Alternatively, footloose carriers such as Ryanair offer a low-cost variant of the same risk.

Recent research (Gillen and Lall, 2004) has suggested that the notion that the primacy of operational efficiency distinguishes the successful low-cost players may be too simplistic. They further suggest that the key lies in the choice of a business model that, as in the case of Southwest, is then complemented by operational efficiency. The situation is further complicated by the fact that low-cost carriers themselves face fixed costs of entry, which are sunk if they then depart.

This risk, in turn, would seem to play an important strategic role in the carrier's strategic choice of a location. The implications of vulnerability have to be faced by airports and, in turn, they need to shape strategies that both reduce risk and allow them at the crucial time of negotiations to strategically influence the outcomes of any agreements with airlines, be they network carriers or their low-cost competitors.

The developing debate on the need for a change in the industry's attitude to innovation forms a significant aspect of the whole question of its ability to introduce structural as opposed to reactive reform. It is suggested, as is noted below, that Southwest's ability to create a formula for its success depended essentially on its ability to adopt a form of 'asymmetric' creativity in which the strategy adopted for growth actually runs counter to the conventional wisdom of the times.

What appears to restrict traditional incumbents in the market when a new entrant appears offering a radically different service is the pull of old habits and assumptions. In the US legacy carriers case, it appears to have been the rather arrogant assumption that the travelling public would simply pay higher charges for what are often marginal services. In fact, when presented with the choice, the public voted with its feet, led by business travellers, tired of absorbing the opportunity costs created by airport delays and the struggle to get from airport to appointment in major cities. The implications of this assumption are investigated briefly below, with another example of what is proposed as a major new force in the competitive market.

Aviation's Future and the Possibilities of Disruptive Innovation

The idea that market economies need to recognize that innovation is both essential and at times very disruptive, was first coined by the famous émigré economist and social scientist, Joseph Schumpeter. He suggested that western capitalism at various periods in its evolution needed to experience what he called 'a gale of creative destruction'. In a very real sense, the aviation industry has been experiencing the destructive part of the process. The question is, are its leaders prepared to think outside the box? This is the clear message of a recent study of innovation theory (Christensen et al., 2004).

Their answer to the comment that headed this section of the chapter is simple: using the famous 'flight or fight' model of both human and animal behaviour. They go on to suggest that low-cost airlines are very unlikely to fundamentally re-shape the airline industry, because 'flight' is not an option for the legacy airlines.

In other words, the only viable choice facing the network companies is to 'fight'. By implication, low-cost carriers can and have captured a significant chunk of the lower end of the passenger market. But any real attempt to move into the rapidly expanding regional markets now favoured by the legacy carriers means that they must abandon the classic short-haul point-to-point system invented by Southwest, in which location, frequency and time are the primary advantages. Regional services will inevitably move them to something much closer to a networked system, with its attendant increase in fixed costs. It is a strategic question already facing Southwest, as it seeks to plot its next growth curve.

A nice example may be found in the case of Jet Blue, which is essentially 'an airline within an airline' (see Figure 2.2 for examples). It offers passengers, attractive amenities with relatively low cost, which may explain why in 2004, its market capitalization is US$2.3billion, about 50 per cent higher than American's. There is some evidence, the authors suggest, that the networks are already moving to what they call overlapping value networks, which may signal growth in a new context and through cooption.

The strategy involved takes the form of a parallel development of a low-cost carrier, operating under its own logo and adding a modicum of services to distinguish it from the low-cost variants. On the basis of the US evidence, such ventures have not been very successful. Where the major legacy carriers have been involved, their attempts to straddle what are in effect separate and competitive markets have not met with a significant response for the travelling public. By contrast, the dedicated low-cost sector has continued to grow its competitive portion of the market, though it is now facing competition from the emergent regional carriers, whose 50-seat turbo prop aircraft have been hailed by some analysts (Christensen et al., 2004) as a new wave in North American aviation.

US: Song (Delta) United (Ted) American (American Eagle) Jet Blue
Australia: Jetstar (Qantas)
New Zealand: Freedom Air
Singapore: Tiger
Malaysia: Air Asia
Scandinavia: SAS (Snowflake)

Figure 2.2   International examples of an airline within an airline
Source: ICAO Journal, *59, 6, 2004, p.16.*

The Vexed Question of the Growth of Strategic Alliances

The tendency for firms to seek strategic advantages through alliances and joint ventures has long been a feature of both domestic and international business activities worldwide. Governmental responses in terms of the law relating to business behaviour have ranged from overt sanction in the German case and through cartels, to downright prohibition in the United States. The last quarter of the twentieth century, saw both the expansion of the MNE and an increasing propensity for mergers, strategic alliances and joint ventures to become the working tools of competitive growth. By the end of 2003, it appears that airlines had become very active participants in this style of business activity, with over 600 recorded commercial agreements. The terms of such arrangements are wide ranging and include those listed in Figure 2.3.

1. Code sharing
2. Marketing cooperation
3. Pricing agreements
4. Inventory control
5. Joint ventures
6. Frequent flyer programmes
7. Blocking agreements
8. Schedule coordination
9. Franchising
10. Facility sharing staff/offices/airports

Figure. 2. 3 Types of commercial agreements involving strategic alliances
Source: ICAO Journal, *56, 6, p.17.*

The evolution of such agreements and their growing complexity as well as consolidation has inevitably raised the questions of the adverse effects on competition and ultimately consumer options. The response of governments and regional regulatory bodies (ICAO, 2004) has tended to see ad hoc decisions of varying regulatory intensity being introduced.

The high-profile alliances contain the most prominent of the legacy carriers, both nationally and internationally. At the current time, the leading organization is clearly the Star Alliance, with 18 members, 15 of whom are mainline carriers. There are indications of a growing maturity, with the strategic direction of the group now subject to consideration by twice-yearly board meetings and a permanent staff of around 70. The emphasis on structure is not shared by One World, which is ranked second and which, according to one major member, is aiming to be a flexible overlay for autonomous members, rather than a global brand. There remains a third alliance, Sky Team, which is the second of the top three and aiming for further membership growth.

There is general agreement that such significant innovations as code sharing have been an important source of revenue, as new city pairs and extra-connectivity raise the level of interline feeds. On the other hand, the original emphasis on increasing revenue is now increasingly balanced with cost saving. The establishment of a fuel purchasing company to service the needs of members has proven very profitable. In turn, its concentration on 'middleware' as a means for linking its members is showing potential for usage growth as a service outside the alliance and even the airline industry.

The steady growth of alliances opens up intriguing possibilities for further competition between the three major groups. The big strategic question is which one will claim the allegiance of China's three large carriers, given the fact that it has now entered into a bilateral agreement with the United States? In addition, the question of future links with India and Russia will clearly emerge at the appropriate time.

What are the Implications for the International Airport Sector?

Within the general pattern of events surrounding the growth of alliances, there lurks the implicit question as to the ways in which the ongoing process of market globalization will treat the airline industry over time. Individual carriers such as Emirates and British Airways have gone on record that globalization is their ultimate aim. In turn, the merger between Air France and KLM would seem to be a possible harbinger of things to come. By contrast, the potential evolutionary model offered by Star may offer a more realistic approach where the demonstrable logic of structure reaches a specific stage not uncommon in international business. This is the tipping point, when the pressure of market demands for an economic strategy based on institutional integration overcomes the national cultures of the various corporate entities that constitute the alliance.

This chapter has attempted to present a series of snapshots of the varied and complex issues now facing the aviation industry. By definition, all of the issues and problems that have been discussed may be found in the airport sector, albeit shaped and formed by the institutional roles prescribed for the many and various types of operations that airports are designed to undertake. It will be the primary intention of the succeeding chapters to focus upon the international airport as a specific operational entity, with the East Asian region as the overall location under analysis. The primary focus will be on the effects of market deregulation, which, following events in 1978 in the United States, spread throughout the global industry, and its effects on the complex tasks that face international airport managers (Vernon-Wortzel and Wortzel, 1999).

Chapter 3

Some Impacts of Deregulation on International Airport Development

Introduction

The focus of discussion in the earlier chapters attempted to address the effects of the forces of deregulation, privatization and liberalization, and on the airline industry in somewhat broad and general terms. The time has now come to assess their impact on the international airport as the primary focus of the book. The basic purpose of the current chapter is to examine some of the major issues now facing airports as major agents of international commerce and business. It will do so as a prelude to further analysis of the activities of the major players in East and Southeast Asia, where competition to maximize national shares of the growth of international traffic, both passengers and cargo, is becoming increasingly intense.

Quite clearly, a successful outcome for any of the potential mega hubs both operational and in the construction phase will ultimately depend on the ways in which management adapts to new market challenges.

The Need to Establish the Role, Scope and Dimensions of Change

It is self-evident that the aviation industry has not been exempt from the geopolitical and economic forces that are consistently re-shaping the nature of economic integration. This means that all sectors of the aviation industry are urgently required to address the matter of structural change, but not only within each functional location. Economic integration on a worldwide basis could not have reached its current state without the development of increasingly sophisticated forms of air transport. The increasing significance of the aviation industry as a major contributor to the growing internationalization of trade and commerce has therefore a shaping influence on the future of international airports.

The central question of the longer-term strategic effects of deregulation and privatization on the modern international airport is by definition a question, first, for the design of systems of corporate governance, and then for the operational and managerial activities that flow on from clearly defined strategic decisions based on policy targets and expected outcomes.

It is important to remember at this juncture, that in the key economies of East and Southeast Asia, aviation policy is often shaped by a national and developmental growth imperative, as well as by a search for commercial and market driven advantages. This is particularly true of those states, such as Singapore, South Korea, the Hong Kong SAR, Malaysia and Thailand, whose geographical locations are an increasing source of comparative advantage, as new aircraft technology overcomes distance.

The Economics of Deregulation, Privatization and Liberalization

The conventional economic logic that has driven international market reform since the 1970s requires that considerable attention be focused upon matters of operational efficiency, as a direct consequence of the deliberate push toward the removal of regulative constraints. The presumption underlying what has become an international strategy is based on the belief that free market competition produces important advantages and optimal results for all participants.

Quality performance over time is measured by such factors as consistent revenue growth and economies of scope and scale, as well as the close management of financial risk and the need for a suitable level of return to investors over time. The signals of success are then summed in a positive balance sheet and an attendant growth of the business. While such rules have consistent value in the operational sense, in the real world they tend to strike complexities of a range and scope that require more industry-specific forms of managerial policy and control.

In the airport industry, for example, corporate management must make strategic provision for those issues that have a geopolitical, environmental or structural context and which create the need for an effective corporate response. The advent of deregulation, which in all industries engendered a search to respond to a market driven competitive environment, finds a current example in the case of international airports. It is reflected currently in the growing competition between national hubs in East and Southeast Asia who enjoy a significant degree of locational advantage and all of whom seek a significant share of the increasing passenger and cargo traffic that is predicted for the Asia-Pacific region.

Most countries face a series of major problems when they pursue this approach. For example, the planned increase of ultra long-haul point-to-point traffic that is currently reflected in the introduction of the VLA type F/6 services by major airlines is a current reflection of a rising demand for a wider range of airline services. But the limitations of existing runway and service capacity at most of the key hubs raise the problems of increasing congestion as an economic externality.

Deregulation also adds a significant dimension to market competition due to the sheer size and multiplicity of forms of congestion noted in the previous chapters and found especially in the land mode of a growing number of mega cities with very large populations. Pudong International Airport in Shanghai, for example, serves an urban population approaching 20 million spread across its metropolitan and regional city network in the Yangtze delta. From an airport perspective, inbound and outbound congestion, especially in gateway hubs, is compounded by problems of location and access after landing. This has prompted city administrations toward the adoption of multimodal forms of transportation as a logical solution.

The Growing Impact of Logistics and Supply Chain Management

The role of air transport as a major and growing force in international economic integration or in its more popular form, globalization, has been acknowledged. It is now important to identify two other major forces that have shaped the process. First, the unilateral departure in 1972 of the United States from the Bretton Woods agreement, which had managed the international monetary system after 1945, had two important consequences. The US dollar lost its status as the universal currency against which all other currencies were pegged. In addition, the link between the value of the US dollar and gold was removed. This meant that all currencies were free to float. It also meant that the value of gold, which had also been pegged, was able to find its own level. As a result, both money and gold became subject to market perceptions of their value. As a result, they have since become a tradable commodity, with a price per troy ounce that is now fixed by market demand.

The second important force to emerge was shaped by the growth of information technology. The ability to move money anywhere on the world's surface and in electronic real time, has led to the metamorphosis of international firms into multi-country networks. As a consequence, production, distribution and sales now take place across several countries, with market information and capital moving in tandem. Such developments have also created growing demands for supply chain and logistics systems that allow for faster and more efficient distribution of goods and services between the producer and the customer, which has very important implications for international hubs. The operational stress in cargo markets for the delivery of manufactured goods is now being placed on closer proximity between production, assembly, distribution and final delivery, with the supply chain getting shorter and shorter in real time.

Time, Distance and Supply Chain Management

Aircraft technology now allows cargo requirements to be increasingly customized by effectively offering carrier clients consistently shorter time periods for the movements of goods. These services are exemplified by the number of airports that offer dedicated cargo facilities and have comparative advantages based upon not only their closeness to specific centres of production, but also their relative proximity to other service modes.

Examples of this type of development are offered by the three airports below, which share close proximity at what can best be described as one of the greatest aviation locations in the world. The PRD is one of the great regional entreports servicing China's economic development with over 31 per cent of national exports passing through its transport systems.

- *Hong Kong International Airport*: Offers multifunctional services for both passengers and cargo. It also maximizes the use of its physical location by offering a marine cargo ferry service between its site at Chek Lap Kok/Lantau Island and some 17 ports within the PRD region.
- *Shenzhen Airport*: Within the same regional location, offers rail terminal connections as well as access to oceanic ports, virtually within the same physical location. It also includes logistic access to the delta's inland waterway system and entry into China's land transport networks.
- *Macau International Airport*: Located on the western arm of the delta, the city enjoys emergent status as the 'Las Vegas' of China. It also offers a range of specialist cargo services for perishable goods as well as animal and equine exports. It is one of five airports within the delta with a dedicated international service capacity.

The network has now been capped in 2004, by the opening of the Baiyun International Airport in Guangzhou, which has the space to expand into a five-runway facility over time. As a later chapter will reveal, it also has the potential to enter into competition with Hong Kong, as the market values of this regional location are maximized.

The Persistent Presence of the State in the Market Deregulation Process

The international market deregulation, which began in the early 1970s, has not produced the results (Friedman, 1999, Hutton, 2002, Mercer, 1999, Soros, 1998, Stiglitz, 2003) predicted either by its supporters or by its critics, whether politicians, international agencies, academics or media gurus. The fact that the internationalization of trade and commerce is an empirical reality is not in contention here. Nor is the fact that it was assisted in no small part by the growth of the aviation industry as a primary mode of transportation and other forces discussed above. What seems to have been largely discredited over time is the popular assumption that the evolutionary logic of market liberalization would inevitably lead to a decline and final departure of the state as an agent in economic affairs. Empirical evidence that this is not the case informs further discussion at this juncture.

In the case of the EU's 'open skies' policy, which introduced an active cabotage strategy involving all its member states and in doing so opened the door to the low-cost carriers, we see a remarkable and underlying strengthening of central authority. This is because the law deregulating the European airlines in 1997 sits firmly within the larger legal definition of the Single Market Act 1992, which treats the member states as members of the world's largest single market.

Some indication of the executive power exercised by the office of the Transport Commission within the EU administration may be found in a very recent reaction by the Commission Vice-President to a failure to comply with legislative requirements. On 17 February 2005, new legislation awarding compensation to passengers who have been subject to flight cancellations, denial of boarding rights and losses incurred by flight delays finally became EU law. It called upon member states to put into administrative place the means to both penalize airlines contravening the law by 'dumping' and to impose the means of payment. Needless-to-say, the law has not been popular either with the airlines or certain state governments.

Following the failure of six member states: Austria, Belgium, Italy, Luxembourg, Malta and Sweden, to both instigate and apply the penalty requirements for infringement, the Commission has now launched legal action for infringement against the recalcitrant states, expressing the view that 'all member states must have effective penalties against airlines which do not meet their obligations'. The fact that the power of the state remains an important element in the management of ostensibly deregulated institutions is further complicated by semantic problems relating to what the terms deregulation, privatization and liberalization actually mean in practice.

Some Semantic Problems with the Language of Market Reform

The popular assumptions that have driven market reform during the last 25 years tend to perceive of market freedom as being brought about by the effective removal and subsequent absence of government from the market place. As a result, important terms, most notably deregulation, privatization and liberalization, have tended to be used synonymously, when in fact they have quite distinct meanings:

- *Deregulation*: Refers to the market as a means of guiding economic activity in such matters, as the allocation of resources.
- *Privatization*: Identifies the firm as the key supplier of goods and services.
- *Liberalization*: A shorter version of the term neo-liberalism, used by economists to describe the model free market economy. It is strongly associated with the neo-classical school of economics at the University of Chicago, which has dominated the thinking of bodies such as the International Monetary Fund (IMF) and World Bank until relatively recently.

The extent of market freedom at this stage of its evolution is a matter of degree, rather than a fully developed operational activity to be found in most if not all national economies. There can be no doubt that a significant amount of market freedom is now a marked feature of aviation services in both the United States and Europe. There remains in most cases, however, a continuing dependence on the role of the state (Emmons, 2000) as the vehicle for both the definition and enforcement of collective and individual rights.

Recent research (Pitelis and Schnell, 2002) on the question of continuing barriers to entry in European civil aviation markets has revealed that despite the passing of the 1997 statute, which introduced total deregulation into Europe, there have been significant problems associated with new players seeking market entry into specific routes. The most important problems identified in a multi-country survey involving

36 airlines were found to be a series of by-products of increasing market competition (see Figure 3.1).

1. A lack of desirable slots for both landing and taking off.
2. The need to enter new routes on a large scale in order to balance costs at an acceptable level.
3. The possibility of retaliation by route incumbents, price cutting or frequency increases.
4. Use by incumbents of hub-and-spoke dominance and code sharing as a barrier to entry.

Figure 3.1   Limitations on new market entrants seeking route entry under Europe's open skies legislation
Source: Pitelis and Schnell, 2002.

It is important to note here that the authors reported that there was a wide perception of the effects of these barriers by different respondents. These ranged from minimal influences on the decision to enter the route to the reason why the idea to do so was finally abandoned.

In a very real sense, such evidence is reflective of the decision by the ICAO member states that the future of market access for the air transport industry should be shaped by processes that would lead to 'gradual and progressive liberalization' (Gunter, 2003) and that the selection of due process should be left to the individual member states. We now turn attention to a working example of a national response to the ICAO strategy, and the somewhat controversial response to its application and practice in an international setting.

The US Strategy on Deregulation: Open Skies and the EU

As the progenitor of airline market deregulation, the US political strategy was to create contestable markets open to new entrants. From a managerial perspective, the intention was to replace the club-style membership of the CAB era. In reality and following the legislation of 1978, the incumbent legacy carriers simply increased their share of the competitive market (Wells and Wensveen, 2004), by a series of takeovers and mergers. By contrast, new entrants tended to 'crash and burn', due to under-capitalization and claims of predatory pricing by the major incumbents. The resultant emergence of fewer but larger major carriers reflected a trend to be found in other industries in what was, at the time, a global trend across world markets.

From an international perspective, the US current model of an efficient market liberalization policy has been seen in the systematic expansion of open skies arrangements in which bilateral country-to-country agreements have been the primary tools. This has led to conflict with the EU and with the WTO, which favours a more comprehensive approach to the EU's proposal for a US–Europe–Atlantic Ocean arrangement.

The EU relationship has seen a considerable growth in tension. The EU's perception of market liberalization is far broader than its American counterpart, which is strongly supportive of ICAO's stress on gradualism. The basic European model would involve the expansion of the existing arrangement whereby a transatlantic version of the European legislation would allow free access on both sides of the ocean not only to ports of entry, but also into and between airports within the geographical hinterland.

By contrast, the American strategy has been to create bilateral agreements with the individual member states of the EU, while stalling on the question of a wider and more reciprocal arrangement on European lines. This run counters (Loy, 1996) the original policy statement of the US Department of Transportation published in 1995. At that time the United States was seeking access not only to 'key hub cities overseas, but also through and beyond them to numerous other cities, mostly in third world countries'.

In reality, the United States has been consistent in its opposition to any form of liberalization (Rhoades, 2003) that would allow foreign carriers freedom of access beyond gateway hubs in the United States. Relations with the EU have now reached a stage where the claim has been made that bilateral agreements between the United States and its member's states, are in fact illegal. This is because they transgress the regulations on international agreements currently found under the Single Market Act 1992.

This claim has now been confirmed by a judgement of the European Court of Justice, which has ruled that bilateral aviation agreements are in direct contradiction of the principles underlying the concept of a single market. It follows that any advantages conferred by the said agreement should apply to all member states in the EU.

The United States and Market Liberalization: Globalization and the WTO

The American view on the application of the WTO GATS is congruent with the ICAO position. In other words, individual states should pursue liberalization at their own pace and through means ranging from bilateralism through regionalism, pluralism and multilateralism.

This runs counter to calls for the limited number of aviation services currently listed in the Annex on Air Transport Services attached to the GATS to be expanded both in number and in occupational classification. The objections raised by the United States appear to be based on the need for progress to be made on a national industry-specific basis, rather than a broad spectrum approach. The view is buttressed by assumption that the latter approach would slow rather than accelerate the liberalization process.

The Role of Airports in International Bilateral Agreements: The US view

From an airport perspective, the proposed strategy enunciated by the United States contains a number of important issues. Of these, those listed in their ICAO Conference statement of 2003 have major importance, as shown in Figure 3.2.

1. Unrestricted route rights (Freedoms 1 through 6) applied to any point in each party's territory.
2. Unrestricted capacities and frequencies, with no restrictions on the number of designations and equivalent access for non-scheduled operations.
3. Unrestricted operational rights, including change of gauge, type of aircraft used co-terminalization and inter-modal rights.
4. Pro-competitive provisions on ground handling, sales operations, and non-discrimination are access to custom services and appropriately developed user fees.

Figure 3.2   ICAO's proposals for the evolution of market deregulation in national aviation industries
Source: ICAO, 2003.

In the vexed matter of airport congestion, a formal call is made for ICAO member states to expand their infrastructural capacities in order to overcome what is seen as a constraint on market access. The approach is somewhat offset by the assertion that congestion has not been a significant constraint on the conclusion of liberalized air services agreements between member states. The implications that a continuing increase in services liberalization might clash with the time requirements for infrastructural expansion, given environmental, structural, spatial and financial constraints, seems to have been missed by the authors of the statement.

The overall image left by the US view of what constitutes an effective open skies strategy appears to cast airports in a supernumerary role. Their primary purpose remains cast in the traditional mould of the supply of efficient airside services, at mutually agreeable cost to the airline as the operational client.

An Alternative Asian View: Japan's Strategic Plan for Airport Development

The United States position stands in marked contrast, for example, to the Japanese belief that, from a national perspective, airports in the East and Southeast Asian regions are growing in both strategic importance and competitive market strength. Government strategy is as a consequence based on geopolitical as well as economic considerations. According to one authority (Hasegawa, 1996), major investment in airports is currently a key development strategy in Japan, because aviation will become in the twenty-first century the most important infrastructural link between Japan and the rest of the world.

This view gains support from further Japanese studies (Yoshikazu, 2000) that have attempted the longitudinal analysis of passenger movements between 34 cities in East Asia, for the target years 1985–1990–1995–2000. The intention was to assess the relative growth of origin and destination traffic (O-D) over the 20-year time cycle. The samples involved those major routes that were serviced by multiple and not single carriers. The macro data reveals that the total network system had increased in size from 54 routes carrying 16 million passengers in 1985 to 117 routes carrying 49 million passengers by 2000.

The key routes reflecting a consistent and rising passenger growth over the survey period included the following important pairs of hubs:

- Hong Kong–Tokyo
- Tokyo–Seoul
- Singapore–Kuala Lumpur
- Singapore–Bangkok.

In contrast to the general trend, Tokyo to Hong Kong actually showed a perceptive downturn in the period 1995 to 2000. Yoshikazu suggests that causal factors may have included the general decline in the national economy of Japan after the bubble economy collapsed, coupled with the major restriction problems on traffic experienced at Narita during that time.

Discussion thus far has reflected the ambiguities and confusions attendant upon the search for appropriate strategies that would maximize the benefits of deregulation, privatization and liberalization, while minimizing its possible negative consequences. The natural approach of much of the popular literature on deregulation and its various formulations has tended to stress the former over the latter. The question that lurks behind such optimistic expectations is inevitably, are there any negative externalities being concealed as the evolution toward market competition continues?

Airport Dominance and Asymmetric Regulation as a De-regulative Consequence

According to a recent study (Turnbull, 1995), the then eight largest carriers operating in the United States controlled 94 per cent of the passenger markets. In addition, under rights contained within their long-term leasing arrangements, they had obtained virtual traffic control over many of the key US airports. An earlier chapter has considered the issues of predatory pricing and slot control. We will now consider the more complex issues that might arise in a given post-deregulative period and endanger the strategic intentions of the airport that is involved.

According to a recent study (Bilotkach, 2004), asymmetric regulation may arise where the process of removal has led to the situation in which different players face different entry and other barriers. In other words, arrangements that existed between the airport and certain parties prior to the regulative change might affect the degree of contestability between all the parties after the changes are in place. Dominance effects, in turn, are often found (Borestein, 1989, 1990, 1991,1993) in the higher charges made to international passengers whose O-D schedules move them between two hubs of a given carrier that enjoys a dominance position.

A study of business travellers (Berry et al., 1996) suggests that the existence of the dominance effect is such that its contribution to an airline's ability to charge higher rates to users actually contributes more in revenue than may be obtained from competitive dominance over a given route. The authors here would argue the case for taking airport dominance effects into account when the proposed format for regulative change is under discussion. The suggestion has also been made (Bilotkach, 2004) that if the changes once introduced produce an asymmetric regulatory regime, a more restrictive approach should be taken towards the rights to be enjoyed by the dominant airline(s).

This proposal does recognize that a degree of inevitability will always be present where, at a specific gateway hub, the national flag carrier has a dominant position because of governmental fiat, rather than any other source of market authority. How airport management attempts to deal with this order of problems and issues will be further discussed in a later chapter.

We now consider an important question, which, given the complexities and ambiguities encountered thus far, may be the most important issue raised in the chapter. It asks, how far should the process of market liberalization go in defining the extent of freedom and the accompanying degree of privatization that should accompany that choice?

What should the Extent of Privatization in International Airports Be?

There is, perhaps, given the very active popularization of the term, a natural tendency to see the process of privatization as the active transfer of public ownership. The implicit assumption here (de Neufville, 1999) is to believe that ownership actually defines control. On the other hand, the nature of air transportation requires forms and procedures of control that are endorsed through international agreement, with safety as a prime example. A leading question in a private market where management is free to create its own policies and procedures, would be, how can compliance with universal requirement common standards be maintained? De Neufville goes on to suggest that a number of options exist in terms of the control process. Their operational characteristics are listed in Figure 3.3.

It is worth returning at this point to the deregulation model proposed earlier (Emmons, 2000) in which the author notes that the freedom of new firms to enter the market and, by definition, the freedom of customers to exercise choices, is really advanced through the imposition of mandatory access requirements on the incumbent firms, who, under the old regime, controlled some important infrastructure or other scarce resource. Obviously this is a zero sum situation for incumbents, who are then moved by circumstances to defend their turf.

1. *Full public control*: Government is the controlling authority and delegates the total management process to the professional civil service through a dedicated department of state. Its principal officers report, in turn, to a designated governmental minister.
2. *Full privatization*: The state asset in question is sold on the market and complete control after purchase is passed to private interests. It then becomes a private limited liability company, plc or corporation, subject to the control of an executive board and its shareholders.
3. *Shared control*: Government sets the policy for control of the entity and sub-contracts the defined management requirements to an appropriate services contractor.
4. *Regulated control*: Partial privatization has taken place, but government, has either maintained a 'golden share' in the equity, as key shareholder, or has reserved the right to invoke and ratify the use of key requirements, such as appropriate rents, user fees, charging caps and rights of access to airport services.

Figure 3.3   The range of options for managerial control of international airports

Source: de Neufville, 1999.

A working example of such a reaction can be found in the important matter of gate access for competitor airlines, where a dominant airline has had major control over available slots. The question of slot control is very germane in the sense that the central role of national flag carriers as the primary users of major hubs is very common in East Asia, which sums to the mutual endorsement of arrangements by flag carrier, government and airport.

On the presumption that deregulation leads to a more competitive market, why should incumbent firms voluntarily offer new entrants access to, say, slots at prime time, or make available other factors that will disadvantage them in future competition? A significant example of incumbent carrier response can be found in the results that emerged from the entry of new carriers into the US interstate market in the period after 1978. In the years up to 1992, some 18 new entries were recorded. Their time in the market ranged from 1 to 12 years, with an average duration of 5.5 years. Four were taken over by the legacy carriers who dominated the market and the primary cause of the exit of the other new carriers was bankruptcy.

The question of choice, as discussion in later chapters will acknowledge, is very strongly influenced by the special nature of the industry. Both airlines and airports share a common concern that efficiency in services be balanced by operational safety regulations that guarantee protection for all client users and staff at all times. The further question arises of how congruent these requirements are with the demands and expectations of the culture of international commercial competition, where the rules of highest returns for lowest cost tend to dominate. Again, one can ask what effect will the growing propensity for large international firms to sub-contract services to centres of lowest cost, and on a global basis, have on the need for the highest standards of quality control to be maintained in the aviation industry.

All of these questions reflect the need to address matters affecting structure from a perspective that allows for the special and unique nature of the roles played by industrial sectors. The steady transformation of international airports in terms of their roles within the various industrial supply chains as generators of economic growth must clearly be factored in to both the realities of policy decision making and the research that is needed to support the total process.

The exploration of the themes contained in this chapter has demonstrated the wide and complex issues that confront the aviation industry in its quest for a more liberal and market-based environment. Serious and important questions remain to be addressed on both the international and national scale of events and developments. Further discussion in later chapters will return to these topics with some frequency and in different contextual settings.

The clear cultural and strategic differences that both unite and divide the various major industrial and national players are compounded by the fact that most major participants in the aviation industry are at various stages of their liberalizing agendas. This inevitably raises a series of geopolitical issues that need to be considered. Despite the various ASEAN and APEC initiatives that have already been started, progress toward a more efficient, deregulated and competitive market remains limited. The causes may be found both within the geopolitical balance of power in East Asia and in the larger regulative standards that are imposed in the aviation industry on an international basis.

They are further compounded by the forces of bilateralism,as exemplified by the strategies that have been adopted by the United States, the EU and individual countries from outside the immediate geographical region, such as New Zealand and Australia. The next chapter will examine in more detail some of the geopolitical issues that have already been canvassed above. The intention is to demonstrate that there is an active correlation between the very significant and growing economic clout enjoyed by the leading states in the East Asian region and the perceptions of market opportunity for the aviation industries of the major states that constitute the global players in international trade and business.

Chapter 4

The Influence of Geopolitical Factors on Major East Asian Hubs

Introduction

There can be no doubt that the future of the aviation industry in East and Southeast Asia is irrevocably linked to the future economic health of the region. In that regard, the region shares with the rest of the Asia Pacific countries, indeed with the entire international economy, a need to develop policies and strategies that will underpin this healthy and growing service sector. At this point, a number off complex issues arise in which national as opposed to regional interests may sometimes be in conflict. It is necessary to examine some of these in detail, prior to turning the focus of the balance of the book to the central issue of the future roles to be played by the region's international airports.

The purpose in doing this is to identify the key drivers underlying the expected and medium- to long-term demand for aviation services in the Asia Pacific. The expected increases in the need for services are, in turn, shaped by the progressive economic development of the various countries across the region. The emergent pattern of inter- as well as intra-country trade will also be influenced by the degree to which the smaller states cluster geopolitically and possibly within the ambit of super economies, such as the United States, Japan and China.

Of equal importance is the need to recognize that the region that constitutes the primary focus for this study is also inhabited by countries that are increasingly engaged in intra-country trade through the medium of bilateral free trade agreements. This means that the concept of the entire region as a free trade block, similar to the former European Economic Community (EEC), is under acute pressure at the current time. As a prelude to further discussion, this chapter will consider the geopolitical issues that have shaped the emergence of the East Asian region as a major location for trade and economic growth, particularly within the current context of the internationalization of trade and business on a global scale.

In doing so, we must also take note of the fact that beginning in 1997, the leading edge economies popularly known as the 'little tigers' or 'dragons' underwent a major financial crisis as the investment bubble fuelled by cheap credit and significant foreign investment burst. Economic recovery has been long drawn out and, in a large sense, is still going on today in some of the countries that were involved.

There is a further need to identify the special characteristics of the region in terms of the initial movement toward cross-border trade, prior to the events of

1997 and following. The conventional western style of negotiations, as reflected in the WTO–GATT procedures have been both long drawn out, as witnessed in the Uruguay Round, and subject to lack of communications between the parties.

As a direct result, the outcomes for all parties have been sub-optimal. This is reflected in the fact that the anticipated Doha Round remains to a large degree in stall mode at the current time. By contrast (Drysdale and Garnaut, 1993), observations in the Western Pacific gave rise to three forms of adaptation to the regional dynamics of market liberalization, according to the various stages that they had reached in their economic development. In the pre-crisis period that saw the growth of ASEAN, most states followed this sequence of strategies:

1. Growing evidence of the beneficial effects of keeping national borders open to trade encouraged a lack of protectionism.
2. The collective experiences of the countries taking part were not subject to active discrimination in trade policy.
3. Further experience of the effects of non-official barriers to trade and the need to reduce them was a major function of trade expansion.

The immediate question arises, where did the aviation industry fit into this process? It would be timely to first locate the industry within the larger context of the growth of aviation as a truly international industry, which is rapidly becoming a form of mass transit.

The Identification of Aviation's Role in a Multi-modal Transport Environment

The key instrumentality of multi-modal transportation as a major group of industries, most active in the growth of the global economy, has gained increasing and international recognition over the last two decades. Within this growing awareness, aviation has been identified as having its own major role, because aircraft technology has enabled cargoes and passengers to be moved further, faster and in increasingly significant numbers and amounts. This has been particularly true for the western countries, led by the United States and Europe, although they are currently trying to advance the cause of the internationalization and market liberalization of the significantly different aviation industry 'open skies' strategies.

Both the United States and the EU have been materially influenced, as mature economies with significant aviation markets of their own, by the seeming potential for the growth of international services into East Asia, which is seen as the next epicentre of aviation market growth. This has led, in turn, to the introduction by both parties of linkages that will permit the extensive growth of inter-city pairing between their major international hubs and strategically placed airports in the leading edge economies of the East Asian region, with access to China as a primary location.

As was noted in an earlier chapter, the EU has created a cabotage-based system since 1997 within its legal and administrative structure. The United States is firmly of the opinion that competitive, efficient and liberalized aviation markets are best attained through bilateralism, as found under current regulations, rather than through multilateral strategies. They are materially assisted in this stance by the fact that the WTO–GATS Annex on Air Transport Services virtually excludes all matters relating to traffic rights. But it must also be recalled that they operate a dual environment, with the domestic sector virtually deregulated, but the international sector subject to material limitations such as limited access to FDI.

The current result of this important conflict over the best way to advance market liberalization has stalled further attempts by the EU to advance the cause of a transatlantic open skies agreement. The situation has also been exacerbated by the fact that the EU Commission has declared the US Department of Transportation's suit of bilateral agreements with its member states to be in breach of EU law relating to the single market.

Important Differences in the Developmental Experiences of the East Asia Region

During economic developments in East Asia, the sub-regions, such as Southeast Asia, have been historically shaped by a wide diversity of cultural, religious and political and economic factors and experiences. While key economies, such as Japan, South Korea, China, Singapore, Malaysia, Thailand and Hong Kong, constitute a leading edge of modernized or modernizing economies, the balance of states in the region are currently at various phases of their economic evolution. In the western Mekong area, further ideological complications arise, given the fact that their governments now range from Communist administrations at one extreme, (Vietnam PR, Laos PDR) to a military junta at the other (Myanmar).

These complexities are further exacerbated by the fact that there are, within some countries of middle size, significant differences still in existence between their internal regions. These are found in the serious disparities with regard to the economic conditions under which some local populations continue to live. As an example, in Vietnam, the north–south division reflects not only different political expectations, but also the level of economic rewards. In Malaysia, the major difference is between east and west (Sabah and Sarawak), with the emphasis upon differences in economic infrastructure and inter-regional income disparities. The same problem arises in Thailand, with urban Bangkok enjoying quite significant advantages over other regions especially the northeast. The pattern is repeated on an enormous scale, with China's western and northern provinces lagging far behind the south and especially the coastal regions.

The Roles of ASEAN and APEC in the Advancement of Multilateral Free Trade

Within the operational context of international business and trade, the WTO, as the operational legatee of the GATT, exists to promote the development of free trade on a multilateral basis. In doing so, its primary role includes the development of multinational agreements to which individual countries may then become signatories. Inevitably, tensions exist where national interests are perceived to be threatened by the subsequent removal of trade or tariff barriers. As a consequence, the progress made by the WTO has been slow in the extreme; for example, it took 13 years to complete the Uruguay Round. The central concept that informs the WTO process is based on building consensus, and this (Schott and Watel, 2000) now appears to have broken down.

The organization's problems appear to be partially a consequence of the sheer size of the organization, which is now approaching 200 countries. The famous meeting in Seattle in December 1999 has been hailed by the left as a triumph over capitalism. In fact, the meeting would have been counter-productive in any case, given the fact that the agenda was incomplete and the decision-making process was already operationally dysfunctional. It must also be borne in mind that these structural problems already existed before the event.

Further complications have also arisen as individual countries develop their own programmes of domestic reform, including unilateral trade liberalization. These activities are evidence of a deepening of the reform process over time and at national level, coupled to a growing awareness of the benefits of intra-country trade. In addition, over the last 20 years or so, there has been a tendency for regional trade blocs to emerge, of which ASEAN is a prime example, recognized by the WTO.

By contrast, APEC includes the United States and China in its 21-country membership, but as the word consultation indicates, does not have the prescribed status or the defined operational functions enjoyed by a fully fledged trade agreement. On the other hand, it does have significant geopolitical clout, since its 21-country membership includes the United States, China and Russia.

The APEC agenda for aviation does provide for significant research-based activities through its Small Group system. Its primary role may be seen as the development of a framework for reform on which various courses of action can be identified and defined. The future role of APEC, does not rule out the possibilities (Findlay, 1997) that it might become an active participant in the reform process, since the group has become an active supporter of free trade and market liberalization. Two activities of note already on the research agenda that might materialize in a practical sense involve the possible introduction of slot auctions for air space, and the establishment of new dispute procedures.

By contrast, the ASEAN organization has developed from an original group of five countries to ten and most recently has formalized relationships with China, Japan and Korea in a new combination referred to as ASEAN10 + 3. It has supported the concept of an ASEAN Free Trade Area, ostensibly in its initial format as a *zollverein*, but now faces a series of problems in maintaining unity on the project. These are reflected in the unilateral actions by individual member states that have progressed and deepened their reform agendas and are now actively engaged in the development of their own bilateral trade agreements, even outside the ASEAN framework. With this qualification in mind, it is now time to consider in somewhat more detail the policy directions that continue to complicate the future of the industry in the East Asian region.

ASEAN'S Approach to a Formal Aviation Policy

The increase in the number of states in ASEAN has also had a multiplier effect on the range and type of airlines in the region, from world-class carriers to small in-country service operators. It has also brought into focus a growing awareness of sub-regional differences. In 1997, ASEAN introduced a formal programme for the integration of a Transportation and Communication Policy. The reason for this decision stems from the fact that, of the member states, only four, Malaysia, Singapore, Thailand and Vietnam, are physically connected. Since the principal driver behind the unification strategy was the need to advance in a positive way the interconnectedness of member states, the addition of air space management to the list of service industries to be considered was both logical and timely.

In the fourth programme of the proposal, the authors of the report extolled the virtues of regional airports and their ability to add value to key industries such as tourism (Proceedings of the Third ASEAN Transport Ministers' Conference, 1997). But at the same time, the alarming effects of growing congestion on some of the region's busiest routes were seen as an important problem. Estimates of the additional cost of congestion for airlines found these to be in excess of US$100million, and the effects on this on tourism as a key regional industry were canvassed.

In turn, the seventh programme called for an open skies agreement in which fifth freedom rights would become common to all member states. The experience of the Philippines's internal deregulation policy, which allowed five new carriers to compete against the national flag carrier, was cited as an operative example of unilateral action by a member state within its own internal market.

In 1995, the Bangkok ASEAN Summit had approved a decision to create a framework agreement on services to be moved forward as a regional activity. The intention was to solicit responses from member states and the GATS framework was used as an operational template. The further purpose was to encourage member states to advance their individual positions on liberalization beyond those they had already taken on the GATS, on the hypothetical assumption that the cause of regional market liberalization could then be advanced.

The Problems of Request and Offer as a Policy Instigation Device for Aviation Reform

The inherent weakness in the approach taken by ASEAN to instigating reform policies can be found in the use of what has been called the 'request and offer approach' (Nikomborirak, 2001). The intention was that individual states could then put forward their views on the market liberalization process and also, in turn, signal the extent and degree to which they would be willing to change their own current practices.

The proposal required individual member states to instigate change unilaterally, and of their own unilateral volition. This project, intended as a preliminary to an ASEAN Framework Agreement on Services (AFAS), covered seven services, including marine transport and aviation. Research on the results sampled five national responses from Indonesia, Malaysia, the Philippines, Singapore and Thailand, all founding members of ASEAN.

It was believed by the instigators of the project that the notion of regional negotiations would elicit much bolder commitments from these countries than those they had put forward in discussion on the GATS. This assumption was further supported by the absence of the broad stream diversity and international participation that surrounded the GATS proposals.

In the original test model (Nikomborirak, 2001), the responses of the ASEAN-5 countries to the GATS framework was used as the basis for negotiation. The states were then required to present their requests for liberalization and, in turn, offer their own proposals for internal reform. It was found after application of the invitation section in this methodology that there was little real interest in the AFAS approach. As a result, an additional requirement was added to the offer section that any proposals should at least advance beyond the commitments made in the GATS proposals.

Further testing (Nikomborirak, 2001) then measured the degree and extent of differences between the GATS and the AFAS data. The result found that the AFAS returns were only marginally better than the GATS. In the case of the air transport evidence, the distribution was interesting, as Table 4.1 reveals.

Table 4.1   The reluctance of ASEAN-5 countries to seek an open skies arrangement

Country	GATS commitment	AFAS commitment
Indonesia	0.55	0.54 (+0.1)
Malaysia	0.25	0.08 (-0.17)
The Philippines	0.85	0.56 (-0.55)
Singapore	0.00	0.12 (+0.12)
Thailand	0.55	0.5 (-0.05)

Source: Nikomborirak (2001).

While the Indonesia commitment appears to be marginally in favour of the GATS, both Malaysia and the Philippines, appear to be to have little interest in AFAS. What is significant from the perspective of the total services group is the fact that the evidence for air transport is replicated in all of the other services sectors.

In their overall interpretation of the data, the researchers found that there was a clear unwillingness on the part of member states to open up their markets within ASEAN. In fact a zero sum game akin to the problems faced by WTO negotiations seems to have shaped the offer response, with the clear expectation that market entry should only proceed on the basis of concessions by every other member of ASEAN-5. Progress also seems to have been impeded by the continued existence of what appeared to be four major weaknesses that would continue to limit any real progress in the liberalization of services throughout the region:

- lack of political will;
- lack of a genuine commitment to open up services markets;
- innate weaknesses in the negotiating framework;
- legal, administrative and institutional limitations.

Free Trade Agreements in the Asia-Pacific Region

Free trade agreements (FTAs) are trading arrangements where two or more economies decide to remove all barriers to trade between them. There can be no doubt that, in the period since 2000, the configuration of trading arrangements in the ASEAN region has changed (Scollay, 2001) somewhat dramatically. While a collective intention to progress the idea of an East Asian trading bloc appears to remain an agenda item, individual member states appear to be actively engaged in making bilateral arrangements for FTAs, both within the ASEAN network and with national partners to be found outside their immediate geographic vicinities. According to a recent study (Rajan and Sen, 2004), FTAs appear to be a reaction to domestic and international constraints. They also tend to exhibit three important features, as shown in Figure 4.1.

Quite clearly, the new order of FTAs are being driven by those economies that have attained middle- or upper-income status, for example, Singapore, Thailand and South Korea. Given the limitations imposed on lower-income developing countries with regard to the serious limitations they face when they enter negotiations with leading-edge states, the current FTA environment in the region might be best described as competitive liberalization.

1. Their structural intentions go far beyond the liberalization of merchandise trade and now include services and other arrangements that contribute to deep integration amongst partners, such as:
    *   investment protection;
    *   harmonization and protection of standards and certification;
    *   protection of intellectual property rights;
    *   opening up of government procurement markets;
    *   harmonization of customs procedures;
    *   development of dispute resolution procedures.
2. They are no longer restricted to their specific geographical regions.
3. As a function of the range and depth of terms and conditions, membership is often restrained either to bilateral or relatively small numbers.

Figure 4.1   The deepening and widening of ASEAN FTAs
Source: Rajan and Sen (2004).

The Political Economy of Current FTA Growth.

The search for the crucial factors underlying the rapid growth of FTAs has to accommodate the fact that while many of them are common to many of the countries in the region, some activities are country specific. Any attempt at a comprehensive searching for, and comparing, country specific activities is really outside the brief of this study. On the other hand, current research (Park et al., 2004) suggests there are five important and common features that can be identified, as shown in Figure 4.2.

1. A contagion effect has been experienced in the region as a consequence of the growth of FTAs, notably in key markets for regional exports. The motivation for East Asia to follow suit can therefore be linked to the strategic need to both maintain and expand market access.
2. The virtual stagnation of multilateral trade liberalization has resulted in the growth of an active bilateralism.
3. FTAs are also being used as a means to stimulate domestic reforms, which had tended to slow down considerably by 2000.
4. Experience has shown that, suitably customized, FTAs can act as vehicles for cooperation and mutual economic and technical assistance between states.
5. The geopolitical rivalry between the major regional powers, China and Japan, for the leadership of the region has led to both countries seeking to strengthen their ties with ASEAN and the newly industrializing nations (NIEs).

Figure 4.2   Variations in the role and purpose of FTAs
Source: Park et al, 2005.

The Emergence of Regional Trade and Preferential Trade Agreements

Both of these categories of agreement have their own special characteristics, which require some degree of definition. A regional trade agreement (RTA) refers to all types of agreements between a small number of countries, with regard to economic policy, who may in fact not share either common or even regional boundaries. By contrast, preferential trade agreements (PTAs) may be described as a sub-group (Findlay and Pangestu, 2001) of arrangements where two or more economies arrange to lower a number of current barriers to trade.

Since 2000, the East Asian region has been experiencing what has been called (Drysdale, 2005) a new regionalism. It has emerged from major issues such as the deep financial crisis of the latter part of the 1990s. This was, in turn, further complicated by the manifest failures of bodies such as the IMF to fully comprehend the structural issues and contagious effects of the collapse of key Asian currencies, as well as the failure of the Seattle meeting to give traction to a new WTO round. The result of this confusion saw the growth of RTAs already been described above, as a response by individual member states to the crisis, especially the leading-edge group, caught by the high debt-to-equity ratios from the extremely rapid growth of foreign direct investment.

It is now clear that by the conventional measure of current exchange rate product values, the East Asian region has some distance to go in terms of geopolitical parity with North America and Europe. But there is clear evidence of the emergence of a significant growth trend, when the same data is measured by purchasing power parity. Data for the period 1991–2002 has ASEAN-10, ASEAN-5 and China–Hong Kong all signalling modest but clear gains, while Japan, the current geopolitical power in the region, saw its export share fall by 2.4 per cent.

There remains, of course, the large question of the emergence of China as the primary economy of the region and its possible role as a natural hub for the region, set within a system of bilateral and preferential arrangements. This assumption brings with it elements of a paradox, since current diplomatic initiatives, led by President Hu Jintao, would seem to point in a global and multilateral direction rather than a purely regional one.

Discussion thus far has been concerned with the economic environments within which the future role of aviation is deeply embedded as a key service sector. The industry has also been given a leading role in the NIEs, such as Vietnam, for example, as a major instrument to be deployed domestically for the purposes of national economic development.

The balance of the chapter will now focus more directly upon the current strategies that are being employed in the regional aviation industry, as a response to changing regional market conditions. Attention will also be given to the fact that such changes are also motivated by the evolving processes of privatization and deregulation, which are going on internationally.

The ASEAN Aviation Industry and its Airline Sector

Before turning full attention in the next chapter to the core questions facing the international airports of the region, which will form the balance of the book, it is important that a review be made of the airline sector and the issues that currently confront its member firms. ASEAN airlines present a quite heterogeneous perspective, as a combination of large operations surrounded by a fringe of smaller firms. Leading carriers, such as SIA, are world class both in terms of profitability and productivity. At the same time, several of the older flag carriers have been in difficulties as the result of various crises over the last 6 years.

At yet another level, a younger generation of carriers, from countries that are relatively new entrants into civil aviation, are struggling to get underway, while a flourishing LCC group, led by Malaysia's Air Asia, are beginning to make their mark. Following the international trend, a wave of regional competitors, some from as far away as India, have signalled their intention to enter the market. The overall impression is of a region where various carriers are at often widely different stages of competitive readiness.

The Current Stress upon Bilateral Agreements and Open Skies

The changing situation in world trade prompted by the current inertia in the WTO with regard to aviation matters, finds countries becoming very active unilaterally in the policy field and on a number of fronts. A most notable example is the United States, which since the passing of the Trade Promotion Act of 2002, has moved into global, regional and bilateral trade negotiations on the grounds that they are all mutually reinforcing activities in advancing the cause of market deregulation. Even before the passing of that legislation, the US Department of Transportation had been quite aggressively pursuing an open skies policy, through bilateral agreements.

Bilateralism really began with the first US–UK air service agreement (ASA), which dates from 1946 and has since been immortalized in the textbooks as Bermuda I (Huenemann and Zhang, 2002). What became Bermuda II was renegotiated at the insistence of the British in 1977, with a much more delimiting set of terms. These stipulated the rights of entry of specified carriers between two countries, the designated airlines permitted to fly the specified routes as well as their carrying capacity.

In 1992, the signing of the US–Netherlands Agreement signalled the commencement of a more liberal regime and began the sequence of further bilateral agreements, which the Americans refer to generically as open skies agreements. In fact, the term as applied to such arrangements can be misleading, since they tend to exclude the possibility of cabotage as an outcome, as well as any extension of the agreement to third parties.

It is also notable (Elek et al., 1998) that one of the most controversial aspects of bilateral-open skies, which are based on the availability of hub-and-spoke services, is the fact that foreign carriers do not conventionally have access to the lucrative domestic US market. An active restriction beyond gateway rights can be imposed by the requirement that a minimum share of passengers on a designated flight is from the point of origin, rather than any intermediate port en route.

As a current working example of some of the negative consequences of market limitations, it is interesting to consider the attempts by SIA to enter the lucrative Australian service to the United States. These have been consistently blocked by Qantas, which currently shares the routes with United. In June 2005, the Australian confirmation of the refusal to change was delivered through a personal message from the Australian Prime Minister to his Singapore counterpart. As a consequence, Jetstar, a Qantas-backed LCC currently trying to build a business out of Changi, has failed to secure landing rights at some premium Asian destinations and has subsequently had to lease out half of its fleet of eight Airbus A320s.

On a more positive note and despite the absence of any major and substantive issues on the liberalization of aviation markets to be found on the GATS agenda, it is worth putting on record the fact that the East Asian region is the site of a somewhat unique multilateral agreement to which the United States is a co-signatory.

The Multilateral Agreement on the Liberalization of Air Transportation, (MALIAT) was negotiated between 31 October and 2 November 2000, at Kona in Hawaii, and signed at Washington, DC on 1 May 2001, coming into force on 21 December 2001. The purpose of the agreement was to promote open skies air services and the initial signatories apart from the United States were Brunei Darussalam, Chile, New Zealand and Singapore. The Protocol to the Agreement provides for the parties to exchange seventh freedom passenger and cabotage rights. The key features are shown in Figure 4.3.

1. An open route schedule to be negotiated between the parties.
2. Open traffic rights as specified and to include cargo.
3. Airline investment provisions that focus on effective control and principle places of business, but protect against flag of convenience carriers.
4. Multiple airline designation, plus third country code-sharing and minimal tariff filing regime.

Figure 4.3   The terms of the MALIAT

Since the Agreement came into force, Samoa and Tonga have become member states and Peru has withdrawn. What is important, however, is the fact that under Article 2 of the Protocol, the parties to the agreement have the right to perform:

• scheduled and charter international air transportation in passenger and combination services between the territory of the Party granting the rights and any point or points; and
• scheduled and charter international air transportation between points in the territory of the party granting the rights.

The agreement remains open to accession by any state that is a party to four major security conventions and has been frequently referred to by senior US Department of Transportation officials as an important model for the liberalization of entire regions. There is also evidence emerging of a widening of the relationship between the individual ASEAN states and the larger Asia-Pacific community, as the following events will attest: in 2003, India agreed to permit ASEAN carriers to operate unlimited flights to Chennai, New Delhi, Mumbai and Calcutta.

Meanwhile, in 2004, Malaysia and Hong Kong committed themselves to a bilateral agreement that allows carriers from both countries to operate passenger and all cargo flights between Hong Kong and any point in Malaysia, without limits on routing, aircraft types or frequencies. In the same year, Malaysia and China agreed on a bilateral arrangement for unlimited passenger and cargo operations.

These events confirm a shift in the region toward unilateral agreements both within and without the region. They also increase the tendency toward a possible multiplicity of arrangements, in which location within and specific RTA is balanced in the case of individual members states, with concomitant arrangements with countries outside the region altogether. One thing appears to be certain from an airport point of view. The current Airport Councils International (ACI) worldwide and regional projections of traffic growth from 2005 to 2020 signal, for the Asia Pacific region, an average annual and unconstrained growth rate of 6.1 per cent in the passenger market and 6.9 per cent in the freight market.

In addition, the introduction of the new categories of very long-haul and very large (VL6) aircraft, whose aim it is to maximize efficiencies through point-to-point mega-hub services, requires key airports to manage such services in tandem with the more conventional hub-and-spoke arrangements. This in turn raises the question of the capital cost of preparation for such contingencies.

The Importance of the City for the Modern International Airport

There can be no doubt that the forces that have shaped western industrial culture, find their spatial identity in the modern city. The evolution of industrial technologies and the adjustment processes that urban societies have made to accommodate them reflect structural changes of a fundamental order. The popular example can be found in the clustering of various activities within operational environments, where proximity is the basic rule. They include (Graham, 2003) e-commerce spaces, passenger airports, fast rail stations, export processing zones and, increasingly, multi-modal logistics enclaves.

The operational conditions for this same pattern of spatial dynamics are now available in the Asia Pacific region and increasingly find their most advanced expression in East Asia. In a real sense, the region enjoys a degree of comparative advantage. Unlike many of the western countries with a history of commercial aviation dating back to the origins of flight, they are not limited in terms of physical location with regard to the capacity to expand and grow facilities.

The potential of world air transport demand is well recognized by the major states of the region and this is reflected in the growth of wide-ranging discussion on the advantages that might accrue from a properly structured open skies policy. At the same time and driven under the terms of specific national initiatives, there is much evidence not only of the expansion of the modernization and expansion of existing major facilities, but also the building of entirely new and multifunctional airports. These are not simply being developed within the context of the service needs of a given urban centre. In fact, they are clearly intended to become competitive players in the race to become the dominant multi-service hub, either at the sub-regional or the regional level of air transportation in East Asia.

This pattern of market competition extended across national boundaries has already been shaping what has been called (Graham, 2003, p. 34) the first wave of adaptation and has taken place at Schipol, Heathrow, Frankfurt and Paris and Copenhagen. The second wave is promised for East Asia, with Hong Kong, Singapore, Seoul-Incheon and Baiyun International in Guangzhou as a major Chinese example of the new format.

There is also evidence of an increasing professionalization of managerial services, which is leading to the further development of companies dedicated to the offer of such comprehensive activities, on a medium- to long-term contract basis. In addition, specific airports are beginning to create their own service companies, who interact directly with airlines, logistics firms and cargo carriers. The pace at which these developments will occur, will of course depend on the basic management culture in any given country in the region. This is particularly so for those countries like China and Vietnam, who have moved from the status of a command economy to a model closer to a mixed-market system.

The central question in China's case, for example, will be concerned with the extent to which the Communist party as the dominant political power will permit market freedom to expand under its own momentum. It follows in the larger regional context that the development of a modern, effective and efficient market for airport services will be directly assisted or restrained by the degree of centrality and market control that remains within the brief of central government or its agents.

It is now time to consider some of the developments related to the growth of urbanization that are taking place in the East Asian region, and in somewhat more detail. The next chapter will introduce discussion of one of the major forces shaping Asian society, and East Asia's citizens in particular. The impact of urban growth on the aviation industry was introduced as a topic in the first chapter, with regard to its impact on congestion. The focus of the discussion will now move towards attempting to discern the massive impact of urbanization on the structure of societies where it is estimated some 200 million people are in the process, over time, of moving from rural to urban locations.

The implications for the location of industrial production and the distribution systems that support it are very significant. They may be considered to be equally important in terms of their global impact, as the economies of the region reinforce their future roles in what is destined to be an economic region as concentrated and significant as any other major trade group in the world.

It is important to distinguish in terms of organizational roles and functions, between ASEAN and APEC. The first group constitutes a formally constituted body, with its own constitutional structure. By contrast, the second group is defined by the letter C for cooperation. This means that body that has members drawn from around the Pacific Rim does not possess an operational mandate in the formal sense of the term.

Chapter 5

Urbanization in East Asia: Its Impact on the Major Regional Airports

Introduction

One of the avowed purposes of this study is to attempt to identify and clarify the increasing economic and strategic significance of the key international airports within the larger contexts of social, developmental and demographic growth in the East Asian region. In effect, this locates the primary purpose and functions of the key airports in a dual position. They are required in the professional sense to service the needs of a growing aviation sector in the leading-edge economies of the region. At the same time, they serve a vital role both nationally and regionally in the provision of essential transportation services, especially in those states where infrastructural development is severely lacking.

In both contexts, they face the need to respond to quite complex factors in the overall growth of demand for aviation services. In doing so, the major airports in the East Asian region especially those servicing a major metropolitan centre, such as the national capital, are finding that the demands of their domestic, international and transcontinental clients are becoming increasingly multifunctional in regard to their effects on the roles and functions of the airport.

One of the major drivers quite common to the region at large can be found in the increasing importance of logistics and supply chain management. East Asia is increasingly becoming one of the most important industrial and financial locations in the global economy. As the processes of production become more widely dispersed in a multi-country pattern, the constant pressures to shorten the route between the producer and the consumer becomes more acute over time. This time-based demand calls for increasingly sophisticated managerial services and facilities. In the current phase of development, they tend to be located either within the spatial dimensions of a given major international airport or quite close by in industrial parks.

A second emergent factor is the increasing importance of a major hub's proximity to specific locations throughout the region, that form the growing networks of international tourism, which is now the largest industry in the world. Modern aircraft technology provides both long-haul and ultra long-haul access to any transit destination anywhere. As a direct consequence, the viability of an airport's location both proximate to and as part of main transcontinental traffic routes, opens up a major role for a key airport as an international destination in its own right, or as a mega-gateway hub where a point-to-point journey may then be converted into a multiple series of hub-and-spoke services.

It will be the primary purpose of this chapter to evaluate both dimensions of these activities, while at the same time allowing for the consideration of a third important issue: the fact that the major international airports in East Asia often have a geopolitical identity, both as practical symbols of successful modernization strategies and as key outlets for national trade in exports. Current examples of this latter development are increasingly found in China and in the new Korean 'winged city' at Incheon International, in the form of adjacent duty free zones. In the latter case, the strategy also calls for a resident community of over 7000 people within the spatial location of the site.

Discussion will begin with a consideration of the geographical and economic space within which international airports in the region are embedded. Many of these are of comparatively recent origin and have a number of further stages of future development, as the example below indicates.

The most recent airport to open for business in East Asia is China's Baiyun International, which is located some 23 km from the important southern city of Guangzhou. It enjoys the enormous advantage of freedom from the restrictions of an existing and earlier operational site, which continue to plague and restrict established metropolitan operators around the world. It opened in July 2004, with two runways in immediate operation and room for at least three more. It is also located in one of China's most important industrial locations, the PRD, as discussion in an earlier chapter has already revealed.

Before beginning to consider the sub-regional dynamics that give physical shape to the contemporary East Asian economies, it is necessary to establish the fact that coterminous with its economic growth, the region is also experiencing a growing form of urbanization. As later commentary will indicate, this is reflected in the continuing growth of cities, some of which are duly recognized as global in their sheer size and influence.

Peripheral Development at the Edge of Metropolitan Centres: The Emergence of a Peri-urbanized East Asia

The term peri-urbanization is used (Webster, 2002) to describe a process where rural areas located on the outer edge of established cities take on the characteristics of an urban community. The key drivers tend to be physical, economic and social and their development is also piecemeal, rather than sequentially planned. The existing community is immediately faced with the need to adjust accordingly. The basic shift requires the need to accommodate both urban and industrial changes.

In doing so, local residents must also adjust not only to a significant level of net in-migration, but also to the emergence of a transitional zone in which the dynamics shown in Figure 5.1 can be identified as ongoing change agents.

1. A shift from an agriculturally based to a manufacturing dominated economy.
2. Employment shifts out of agriculture and into factory and other forms of industrial employment.
3. Rapid population growth, which is often underestimated in official data analysis.
4. Rapidly changing spatial development patterns which are coupled to an influx of investment and rapidly rising land costs, leading to a patchwork form of mixed land use.

Figure 5.1   The peri-urbanization process
Source: Webster, 2002.

The main explanation for the phenomenon can be found in the fact that with FDI as a primary driver, the need for relatively easy access to the kinds of services only to be found in a large city with adjacent transportation services becomes essential. These requirements seem to be compounded by the fact that the MNEs involved in given locations may be operating a regional headquarters in an adjacent country, which makes access to a strategically placed international airport something of a necessity.

It is also important to note before proceeding further that the quality of FDI entering into a peri-urban location is vital. Short-run production requirements largely to reduce labour costs tend to leave derelict buildings behind after the orders have been filled. From a human resource perspective, they also tend to leave workers whose skills are either not transferable or over-supplied within the ambit of choice.

The development of peri-urbanization is reflective of the importance of spatial and economic geography on international airport development. It is now time to expand the discussion to include the other dimensional influences of the economic geography of the East Asian region.

East Asia, International Airports and the Influences of Economic Geography on Transportation

It is a fundamental premise of economic geography that economic activity is not randomly distributed across space, because as empirical evidence indicates, people tend to cluster in locations (Brakman et al., 2001), which then allows for market relationships to become established. Over time, it is suggested, the processes of urbanization, in which cities play a central role, tend to demarcate specific activities in terms of specific spatial areas. The patterns of settlement are thus shaped by the processes of economic and demographic growth, which, in turn, give rise to the development of transportation systems as cities grow and expand their boundaries.

A useful working definition of transportation can be used at this point (Hoyle and Smith, 1998) that suggests that transportation as an activity can be described as the epitome of a complex relationship between the physical environment, the range of political and social activities, and the specific levels of economic activity that are all taking place simultaneously, either nationally or regionally. For the immediate purposes of this discussion, some assistance in the identification of specific locational activities can be found in the classification of the various sub-regions (Paez, 2001) that make up East Asia.

The following categories, described as sub-regions, economic zones and growth triangles, reflect the various range of natural endowments as well as the developmental stages reached by the countries that are subsumed in the list below. While the distinction between sub-regions and economic zones tend to blur somewhat in specific cases, the emergence of growth triangles in the late 1980s marked a distinctly new approach to integration. In their most basic form, they constitute a set of localized arrangements between two or more countries (Thant et al, 1994), in which the spatial distribution of the triangle is aimed at the exploitation of specific, contiguous and complementary advantages in the localized geography of the total area. Their development is dependent upon a balanced relationship between the private and the public sectors, with investment capital flowing from corporate sources and the public sector contributing infrastructural development, fiscal incentives and an appropriate administrative framework.

The East Asian Sub-Regions: The Significance of Demography and Urbanization

East and Southeast Asia as a physical location refers directly to the countries that stretch from the Korean Peninsula to Indonesia. They include in their grouping members ranging from economic superpowers, newly industrialized countries and a number of states that are only marginally capable of retaining a subsistence level of economic growth at this current time in their developmental cycles.

The various geographical locations that can be identified as sub-regions are to be found in the sequence shown in Figure 5.2, which lists countries from the northeast to the southwest. The entire group is reflective of a wide range of ethnic, historical, geopolitical, migratory and religious identities, which in a very real sense identifies the region, from a transportation perspective, as one of the world's most important geographical locations. There are also indicators of a colonial past in which British, Dutch and French and even, in the Philippines case, American influences remain post independence.

While Southeast Asia is subsumed within the larger context of East Asia for the purpose of further discussion, it is also necessary to recall that the entire region is within the larger context of Asia-Pacific, often referred to as the Western Pacific. It remains to note that while the Korean Peninsula is a useful gateway to the region, it is also contiguous with China in the northeast.

1. Northeast sub-region: Japan and the northeast.
2. The Yellow Sea sub-region: North and South Korea.
3. South China Economic Zone: Taiwan, Hong Kong coastal provinces.
4. BIMP East ASEAN growth area: The Philippines and Malaysia.
5. SIJIORI Economic Zone: Singapore, Malaysia, the IMT growth triangle and northern Malaysia.
6. Mekong Economic Zone: Burma, Vietnam, Laos PDR, Thailand, Cambodia.

Figure 5.2   The sub-regions of East Asia
Source: Paez, 2001.

While the general focus on the growth of demand for aviation services centres on the Asia-Pacific region as a whole, there can be no doubt that demographically, apart from the west coast of North America, the East Asian region will remain an important centre on its own. This is somewhat obvious, given its important and continuing role as a multi-modal transportation junction strategically placed between the two hemispheres. But there is another dimension to the development of the region that is best exemplified by the global trends toward urbanization and the growth of global or world cities.

The growth of East Asian cities within the context of what has been called (Smith and Timberlake, 2002) an evolving and global urban hierarchy has been remarkable, if only for the fact that for most of the twentieth century, East Asia was one of the least urbanized parts of the world. Several scholars have suggested that there is a world city system now evident (Keeling, 1995; Sassen, 1998) and air travel linkages are effectively the sinews that link the system together.

In the case of East Asia, this argument gains support from a recent study (Smith and Timberlake, 2002), which quantified global air traffic flows between 100 major city pairs, in the period 1977–97. Their evidence reveals that five of the top 13 cities that they ranked are East Asian: Tokyo, Hong Kong, Singapore, Bangkok and Seoul. A further five are European based, while the balance of three is found in the United States.

The scores on network prominence, however, found that the East Asian group ranked below the European five. The top sequence listed London, Paris, Frankfurt and New York, in that order. How far these figures are influenced by the fact that many of the leading airports in East Asia are in a progressive growth cycle that includes considerable physical expansion, for example, Incheon International, Hong Kong Chek Lap Kok and Guangzhou New Baiyun International, is hard to quantify at this time. For example, in Japan, the deliberate use of multi-nodal links to ease the congestion and overcrowding at Narita, tends to obscure to some extent the amount of actual traffic that is scheduled into that gateway hub.

It is also important to include in this analysis, the fact that the growth of cities in East Asia is also reflective of the major shifts taking place in global urbanization. Estimates indicate (UN, 2000) that that while urbanization is growing on a global scale, a significant shift is also taking place in the distribution of the urban population away from the developed world and toward the developing world. The logical outcome from an aviation perspective of what has become increasingly recognized as a long-term trend can be found in the enthusiasm with which LCCs have attracted customers since the pioneer entry of Air Asia.

Table 5.1   The distribution of urban growth, 2000–30

Year	Developed world (%)	Developing world (%)
2000	76.1	40.5
2010	26.0	73.0
2020	22.0	77.0
2030	19.0	80.0

Note: Urbanization in 2000 sees the developed world with only 31.0 per cent of the total urban population of 2.89 billion. The balance of 61.0 per cent is found in the developing world.
Source: UN, 2000.

These data to some extent indicate that the infrastructure of demand projections for future air transportation in East Asia is actually being embedded in urban growth, especially in the major cities. A degree of caution is needed, however, since the data only signals the potential for passenger traffic to increase. The question of cargo services now needs to be addressed, given the fact that East Asia is becoming, in its own right, a major industrial region within the international development of industry on a world wide basis.

There is also a need to be aware that the economic benefits of urbanization in any given country are not distributed equally across the whole of society. In China, for example, the southern and coastal provinces have benefited throughout their considerable history from the fruits of external trade. More recently, they have been the major beneficiaries from the benefits of the market reforms introduced by Deng Xiaoping in the 1980s and that reached a high point recently (Cartier, 2001; Panitchpakdi and Clifford, 2002) with China's entry as a full member into the WTO.

By contrast, the populations in the more remote and less-developed regions of China, especially in the West, are still waiting to be brought fully into the mainstream of economic growth. The resultant tendency for such populations to migrate to the city in search of work and subsistence has become a common experience in Asia, but it is a truly international phenomenon, since it is also now being widely experienced in the West, including the United States and Europe.

The direction of discussion thus far has been directed toward establishing some of the fundamental issues that confront the East Asian airports as they develop their individual responses to a growing, international and increasingly complex consumer market. It is now time to identify the increasingly multifunctional aspects of the role they are expected to play.

Increasing Market Opportunities and Stronger Market Competition: Some of the Strategic Issues Facing the East Asian Metropolitan Airports

It is important before attention is fully turned towards the matters found in the title of this section to observe that the dimensions of market opportunities and market competition extend outside the geographical periphery of the East Asian region. With this in mind, the primary focus of analysis will be upon those important East Asian players in the international airport industry who cast a considerable shadow in the rest of the aviation world.

A careful check across the regional data reveals that some six core metropolitan airports fit the necessary criteria of an operational identity as an existing and important operational hub. The locations will therefore be Tokyo, Seoul/Incheon, Hong Kong, Bangkok and Singapore, whose ranking as major operators has already met the serious tests of international comparison. They also qualify on the basis of the fact that they enjoy the strategic advantage of close proximity to those major cities that are increasingly identified as global in their functional roles and wide span of influence in the international economy.

Mega Regions and Development Corridors: The East and Southeast Asian Example

The evolution of multinational firms, often through the processes of takeover and merger, has created network structures where design, production, distribution and sales can be carried out in a number of varieties of countries. This form of development continues, supported by international financial markets and information technology that allows managerial decisions to be made in real time, anywhere on the planet.

Modern transport technology, which includes bulk maritime carriers, fast rail services, road networks and air cargo, has a primary role supporting this development. As a consequence, the interaction of urban centres, firms and a new international division of labour is in the process of redrawing the spatial distribution of industries across national boundaries, and increasingly over transcontinental distances. Reliance upon conventional groupings of major cities within a single region is being replaced by networks that bring into play infrastructural endowments (Rimmer, 1994, p. 200) that include knowledge-based university and research institute facilities and important cultural factors, which are linked, in turn, by high levels of accessibility.

The vital importance of some optimal level of population, to be found in key locations within the development corridor, requires no rehearsal here. It holds for the East Asian region as a whole and even if the sub-division between east and southeast is maintained. If, following the UN categorization, a major urban centre has a population in excess of two million, then the East Asian region has 11 locations that qualify: Beijing, Seoul, Pusan, Shanghai, Taipei, Hong Kong, Manila, Bangkok, Kuala Lumpur, Singapore and Jakarta. This allows for a net sub-regional division into six locations in East Asia and a further five in Southeast Asia.

In the primary matter of commodity movements, transportation and communications immediately come to the fore. The classification of certain urban centres as mega cities depends on their ability to act as pivotal agents between one development corridor and another. Tokyo, for example, operates as an important interchange between the national hub airports in the region and the American corridor, which has its location in California. In addition, both Bangkok and Singapore serve as pivotal locations for Europe and the Middle East.

With air traffic in the Asia Pacific region estimated to double by 2040, a number of predictable issues are bound to arise to dominate airport development and construction. They will include matters relating to passengers and cargo as well as changes in airport technology. The need for more runways will continue to need attention, as will the need to allow terminals to interface more smoothly with the increasing range of landside activities in which they are engaged. The matter of security will also increasingly influence the design of terminals, as will the increasing number of environmental concerns that accompany the growth of passenger and cargo traffic. All of these issues will continue to confront airport management, especially in the major hubs.

For the moment, however, there is a need to further identify and describe a shortlist of key players that have been selected as exemplars of the current East Asian list of major hubs. Some initial stress will then be given to the operational aspects of the multiple roles that they are playing as key hubs. The chapter will then finally identify some of the emergent topics that will be addressed in the balance of the book.

Key Airport 1: The Tokyo Metropolis Airports (2)

Metropolitan Tokyo is one of the most heavily urbanized areas of the world. Some 40 million people live in the region, which contains 30 per cent of the national population. The shores of Tokyo Bay are also the homes of the major cities, Yokohama, Kawasaki and Chiba. While Tokyo is Japan's largest city, technically it does not exist. The city was abolished as a legal entity in 1943 in government reorganization; with the merger of both cities and districts, the city became Tokyo Metropolis.

Metropolitan Tokyo is served by two international airports and five local airports. The vast bulk of international operations are based at Narita, a city some 60+ km east of Tokyo, while Haneda, the original main location, is southwest of the city, but is situated nearby in Tokyo Bay. Haneda completed an airport enlargement programme in 2003, and is counted as one of the world's largest operators of domestic traffic. It opened a second terminal in 2004 and has a third runway as well as a third terminal under construction. While it does operate international charter flights, the scheduled services account for four flights per day to Seoul Gimpo Airport, by ANA, Asiana, Japan Airlines and Korean Air.

Narita, which was reaching its operational maximum at the turn of the century, now has a second runway and is currently expanding its international services. The five local airports within the metropolitan area add strength to domestic services, with four of them located on islands in Tokyo Bay.

Tokyo has an added advantage in the fact that is close to major deep water ports at Yokohama and Kawasaki. Both are major logistics centres, and have also been designated as foreign access zones (FAZs) since 1992. As a consequence, major domestic investment has taken place in terminal and related facilities. A further benefit accrues from the levels of accessibility available to travellers throughout the region through the medium of highly concentrated forms of surface transportation.

These entire factors sum to a highly integrated transportation system, ready and available to service those workers who trek to the city every, where 80 per cent of the region's jobs are located. It also adds value to the totality of factors that makes Metropolitan Tokyo a global city and, in addition, a multi-modal entrecote for the industrial companies located in the region. It is also important to remember that in Japan, Tokyo is the epicentre of air transport, which translates into other regional airports of relatively much smaller size.

It is notable that in Asia, air transport of passengers has a different structure to the rest of the world. Use is made of large aircraft, on heavily serviced routes, with much smaller feeder traffic involved. As an example (Schaafsma, 2003), an average passenger load in Japan can be as high as 220 passengers for flights that would be serviced in the United States with aircraft designed as regional shuttles.

Key Airport 2: The Strategic Advantages of Chek Lap Kok Hong Kong's International Airport

At the risk of some repetition of earlier commentary, is necessary to identify the primary features of an airport that enjoys international status as the leading site for cargo services. Hong Kong has a somewhat unique political identity in the sense that since its reversion to status as a geographical part of China in 1997, it has been designated as a Special Autonomous Region. Under the terms of its cessation by the British, it retains a relatively high degree of autonomy as an important economic centre.

Hong Kong is located in the estuary of the PRD, which places it strategically; within a region that is one of the most important economic areas of China. Some 40 per cent of exports move through the airports and maritime centres of the estuary, which makes the PRD a very important contributor to the net economic growth of China. In addition, and from an aviation perspective, its geographical location places it, in the words of the famous quotation, '6 hours from 50 per cent of the world's population'.

Apart from the fact that Chek Lap Kok is located on Lantau Island, which was literally constructed from the seabed as a customized site, it shares very close physical as well as air space with three other airports: Shenzhen International, Macau International and Guangzhou's Baiyun International Airport, which opened in July 2004, is situated some 23 kilometres from the old city site and already has two runways operational and space for five. In addition, Zhuhai Airport is located very close to Macau on the western arm of the delta. Despite a significant investment in its modernization, it has tended to be overshadowed both in size and proximity by the larger players in the delta.

The question of major competition between Chek Lap Kok and the intended mega hub at Baiyun has tended to be somewhat modified by the fact that all five airports have created the A5 Forum for the development of various forms of cooperation in areas such as flight diversions, safety strategies, training and development, and responses to health problems such as SARS. All of these concerns are now subject to ongoing study by designated task forces.

The Hong Kong Airport Authority has been heartened by the growing relaxation of restraints on traffic to the mainland. On 1 August 2004, additional route capacity was granted by Beijing, allowing flights originating in Hong Kong to service some 12 mainland cities. The local management sees this as a step toward the building of air bridges between Hong Kong and China's main cities.

As in the case of Tokyo, Hong Kong International Airport is serviced by a dedicated fast light rail facility that allows for all pre-flight arrangements to be concluded at Kowloon station in the city. It is also beginning to grow around its periphery those business and service operations that are so marked a feature of the modern Asian airport. In terms of the future, much will depend upon the degree to which the A5 Forum can succeed in playing a key role in the building of an economic super-zone, both in terms of the Guangzhou–Hong Kong axis and with the other coastal cities and provinces that lie between Hong Kong in the southwest and Shanghai in the northeast.

Key Airport 3: Seoul-Incheon International Airport: the Winged City of the Northeast

The entirely new international airport at Incheon has a number of unique characteristics that reflect the fact that it was deliberately designed as a future mega hub. The clear intention from a strategic perspective is to grow into a role not simply as a national hub, but also as a major international facility able to dominate the Asian northeast. First opened in 2001, the managerial authority has publicized the new player as more that an airport and for the reasons outlined below.

Seoul's Incheon International Airport, is located on an island created for the purpose and has taken over from Gimpo International as the key international terminal for South Korea. It is also a growing centre with a new town (60,000 houses in the first phase) located in close proximity. The location offers a wide range of business facilities, as well as a free trade zone to incoming business. The plan is to offer the total range of economic services that require international air transport on a global basis. In addition, the island has a range of parks, hotels and resort facilities that are designed for tourists.

What is happening here (Schaafsma, 2003, p. 34) is a change in the very basis of the airport's identity as an industrial company, duly recognized in law. The traditional focus of airports has always been on operational efficiencies, appropriate aircraft technologies, traffic volume and basic airside layout and architecture. By contrast, modern airports like Seoul-Incheon are orientated to business-to-consumer (B2C) priorities, profitability, and share of value and business design that does not treat landside operations as a secondary source of income (see Figure 5.3).

1. International Business
2. Free Trade/Custom Free Business Zone
3. Logistics Distribution Centre
4. High Technology Research & Development Centre
5. Tourism and Leisure Centre
6. Supply Chain Key Location
7. Major Hub Choice of International Cargo Airlines

Figure 5.3   The conceptual forms of activities in an airport city
Source: Park and Kwon, 2003.

It has been suggested (Kasarda, 2000a) that international airports of this scale and magnitude, located as they are, if not in proximity, but in the ambit of major metropolitan urban centres, are really a new form of business environment that he calls an 'aerotropolis'. He goes on to suggest that the key driver of urbanization in such a location is in fact the airport itself. As a final comment before turning attention to the next airport on the comparative list, it is important to note that Kasarda sees the growth of aerotropoli, as an intermediate stage in the development of more futuristic forms of transportation networks.

Key Airport 4: Changi International: Singapore's Asian Doorway

Singapore's international airport at Changi has long been a model of development. Since its opening in 1981, it has contributed an annual return in the order of 9+ per cent to the city-state's gross domestic product (GDP). In addition, it has been a major symbolic corner stone in the evolution of Singapore as a developed economy. Equipped with three runways, three terminals and a fourth low-cost terminal that will service the national flag carrier's new LCC, Tiger, the airport services over 80 transcontinental and regional domestic services. Singapore International Airlines (SIA) will also be the first carrier to offer services on the Airbus A380, (VLA6 class), since it was the first international airline to place an order for the aircraft.

In the quarter century since it opened, Changi has gained global recognition as a major hub in Southeast Asia, located strategically where it has access to both east–west and north–south flows. It is also able to offer transit freight services and, in keeping with its great rival Hong Kong, it has an important maritime link with Singapore's Centre-Port.

In more recent times, Singapore's status as the third largest aviation hub in Asia has been claimed by Bangkok, and it currently ranks fourth behind Haneda, Chek Lap Kok and Don Muang. It also sees itself threatened by the rapid growth, both in numbers and popularity, of LCCs and the successful prototype, Air Asia, is symbolically located just across the bay at the new airport and distribution centre of Senai in Malaysia's most southerly state of Johore Bharu.

While SIA is successfully developing ultra long-haul non-stop services to both New York and Los Angeles, utilizing the Airbus A340–500 series aircraft in a premium business/economy class configuration and it has clearly major plans for the A380, it is also acutely aware that when the A380 increasingly comes into service with the major transcontinental carriers, the economies of scale, particularly relating to fuel burn, might make bypassing Changi an important route-planning option for some airlines. This possibility became a reality in December 2003, when Emirates, which is currently building a global network, began an A340–500 service direct to Australia from Dubai.

Changi is also experiencing strong pressures from both Thailand's Don Muang International at Bangkok and Malaysia's national hub near Kuala Lumpur. The perceived threat comes from the fact that both airports are discounting on service fees. According to IATA, Kuala Lumpur International Airport (KLIA) charges some 28 per cent less for turning around a Boeing 747–400, than Changi, an indicator of the intense pressure on that airport to increase its revenues. In turn, at Don Muang, a more modest differential of 5 per cent is incurred.

The possibilities of a diminished role for the airport is further increased by the fact that the land mass of Singapore has reached a degree of finite limitation. Technical limitations now prevent the recovery of land from the sea, which does not allow for the physical expansion of Changi. In fact, the policy has been a major source of contention between Singapore and Malaysia, which has led to periods of often heightened tension in the last few years.

The current strategy in response to perceived threats is intended to cut costs at Changi by between 10 and 15 per cent. The expectation of a diminished role for the airport is fully recognized; to what extent and degree this will reshape a new role for Changi in the future is hard to visualize. The answer may lie in the inherent possibilities of an aviation RTA that would accommodate an appropriate recognition of the possible benefits of greater integration of services at a regional as opposed to a purely national level. Meanwhile, major airlines continue to use the facility and it remains a highly efficient and well-run international airport.

Key Airport 5: Don Muang International, Bangkok's Historical Hub

It is interesting to find that Don Muang's role as an airport commenced in 1914 with the first training flights of the Royal Thai Air Force. By comparison, in 2004, what is now a very overcrowded site serviced more than 80 airlines, over 30 million passengers, 160,000 flights and 700,000 tonnes of cargo. What is remarkable about this achievement is the fact that Don Muang has been able to achieve these levels of throughput despite problems of location, pressure on space and major difficulties with ground access, as the population of Bangkok has expanded. In effect, the airport has been surrounded over time by suburban growth and is now contained within the city itself.

These problems have been further exacerbated by the fact that the airport has only one runway and in addition retains its original status as an air force facility. This means that scheduling must take account at all times of the shared needs of the military. All of these factors in sum have pointed for a very long time to the need to develop an alternative site for a new airport. Political and other reasons have intruded on the project, but it now appears that a new airport will become operational with an opening date that always seems to be some distance in the future.

Key Airport 6: Suvarnabhumi International Airport, Bangkok (SIBA)

Located in Samut Prakarn province some 30 km east of the city, SIBA's future plan calls for progressive development that will allow for an ultimate limit on traffic, in the order of 100 million passengers per annum, based on four operational runways. The initial operational phase will accommodate 30 million passengers. Two runways will be immediately operational and will allow for 76 flights per hour, as well as the additional transit of 1.46 million tonnes of cargo on an annual basis.

It is proposed that SIBA will have its own electrical railway system connected to a downtown terminal. The core terminal complex on the airport site will have some seven floors and a basement. This will allow for a total floor area of 500,000 m^2, which will make it the world's largest. Plans also allow for a system of new roads, which will link SIBA to the ring road system around Bangkok.

With the opening of SIBA, all international services currently accommodated in the two terminals at Don Muang will be transferred to the new location. After due process, it will then revert to a domestic services-only configuration. From a regional and competitive perspective, the development of SIBA gives Thailand a useful advantage in the anticipated increase in competition for both passenger and freight traffic in the future. SIBA also offers a working example of the forces that are shaping peri-urbanization when it is recalled that its parent, Don Muang was originally located in the suburbs of Bangkok.

It is safe to say that the analysis and management of the competitive advantage of international airports in the East Asian corridor is in its very early stages. Modelling some standard procedures for use as an analytical tool is also rendered somewhat difficult. For example, the Porter five forces model, when applied to a given national industry (Porter, 1990), is predicated on the assumption that the ultimate leader will emerge logically from competition within the internal market. As the evidence from current positions of the five international airports has already indicated, the development of a major international airport can result in the deliberate creation of single monopolist, strongly identified with the national government as prime mover.

At this point, the managing agency can be either a government department or alternatively one of the growing numbers of competing specialist firms that will offer a complete managerial service under contract. Finally, a given airport's perspective on competitive advantage, might well rest on the degree and extent to which its ownership has been privatized or liberalized, with private capital forming the basis of operational, productivity and profit outcomes. All of these variables quite clearly act to shape the operational culture, and these will vary from location to location in a competitive market for international as well as domestic clients.

Given the present state of knowledge, and setting all of these caveats aside, there are a number of practical uses that stem from the original five forces model. In other words, there is a practical and possibly testable relationship (Park, 2003) between the various aspects of the model and the realities of operational practice in what is clearly an emergent competitive market.

Park addresses the need for suitable identifiers for those strategic factors that can allow airport managers to assess the success of their operational and developmental programmes in competition with the other major hubs in the region. The intention is to allow management to think across a number of dimensions, all of which find common purpose in the overall activities of the international airport which involve, space, demand, managerial skills and a clear control over services.

Such a strategy can be found in the example of Changi International, which integrates multiple functions in a highly efficient network. The most obvious example is the engineering company within the group. Apart from its central role in fleet maintenance, it is also able to offer competitive services to other airlines, under contract. There is also growing evidence that major efforts are being made by the group to create a range of activities that will allow the in-transit passenger time to relax and even pay a short visit to the city with tour guidance.

The modified classification based by Park on the Porter model is demonstrated in the list shown in Figure 5.4.

It is clear from earlier discussion that Porter's model is now also required to accommodate the spatial distribution of activities, which may extend over a very considerable area, while encompassing in the process a number of localities. Sourcing from a distance is now commonplace in international business, with the organizational clusters of needed components (Porter, 1998) as the primary locations for service. At primary issue is the shortening of the supply chain to include 'virtual' as well as real suppliers as important links.

Quite clearly, integrated forms of multi-modal transportation service become a necessity, with efficiency measured as the distance between the tailgate of a sea, road or rail service and the waiting hold of a cargo aircraft. The growing evidence of the widening and deepening of the primary role of international airports, beyond the traditional role of servicing the airside requirements of its clientele, raises a number of important questions. These relate to the fundamental influence of important trends within the nature of international business that are now intent on shaping the very basis of the firm–client relationship on a global scale.

Spatial factors

> Airport location
> Economic conditions
> Stage of developmental cycle
> Environmental issues

Demand factors

> O-D demand
> HtS availability

Facility factors

> Range and level of facility standards
> Potential for further expansion
> Quality of physical conditions

Managerial factors

> Senior management strategic and planning skills
> Sophistication of financial controls
> Access to sources of new capital

Service factors

> Level of services/operational systems

Figure 5.4   International airport competitive advantage using the Porter five forces model
Source: Park, 2003.

This requires an attempt in the next chapter to examine the developing importance of logistics and supply chain management. It will include a consideration of the major components in any plan to expand further into overseas markets. It will also consider those factors that an industrial firm with an international client base must now factor into any development plans. Dominating further discussion in the following chapter will be the need to identify those servicing facilities that might be obtained by a firm locating either within a major international airport's cluster, or in some congruent supply chain pattern that suits its specific needs, for a fast and efficient real-time service.

The need to plan locations carefully, with a view to attracting traffic, is exemplified by the case of Zhuhai Airport in the PRD. Prior to the return of Hong Kong to the mainland in 1997, the site was expensively upgraded to take larger aircraft. It now finds itself wedged between Macao International and Chek Lap Kok and handles domestic traffic only.

The problem was compounded by the fact that access to any point in the PRD was available through a number of tri-modal combinations, both externally and from the mainland. It is interesting to find, as the next chapter reveals, that within the context of a larger and highly integrated regional concept, Zhuhai has a second chance, if only in the abstract to fulfil a more dynamic and central role.

Chapter 6

Forces Shaping a Multi-modal Future for ASEAN Airports

Introduction

Attention will now turn to the consideration of some of the primary change elements in East Asia that are having a very powerful impact on the role of aviation in general and airports in particular. They range from the macro shifts that are transforming industrial and urban spaces in the region to the growing importance of major airports as essential agents in the transformation process. While the primary task of facilitating the efficient usage of client services for passengers and cargo as they pass through the airport remains at the core of its operational dynamics, cognisance must now be taken of major new influences shaping the spatial dynamics of the region.

The major hub airports increasingly play very substantive roles in the overall and multi-modal transportation networks that support increasing urbanization in East Asia. In the growth process, the spatial distribution of both logistics and supply chain operations have progressively moved from the immediate physical vicinity of the airport, as demand for services has grown. They are now to be found in strategic locations that serve multi-modal transport flows, and are often located in customized distribution and specialist service centres at some appropriate location, often in a dedicated services park, run by major transport logistics firms.

Contemporary research is producing evidence of airports as contributors to the processes of global interdependence through transportation networks. What is widely regarded (Douglas, 1998) as a major trend in spatial transformation has had consequences that are still in the process of rapid evolution They include forms of profound structural change, such as world city formation, urban restructuring, transborder regional development, the creation of coastal metropolitan regions and the emergence of mega-urban regions, as well as the peri-urban development discussed in the previous chapter.

Underpinning these developments (Van Londen and de Ruijter, 2003) is a massive exchange of people, goods, services, ideas and images, all connected on a global basis by telecommunications and transport technology. As a direct consequence, more and more people are drawn into the exchange process, while remaining physically at home within their local communities.

Developments of such a range and scale inevitably raise the question of the costs incurred and the need to obtain financial support. While governments have been the traditional source of such revenues, the role of FDI has become central to the entire process. Two further and major influences now shape national strategies in the case of airport development. The first, as discussion above has already observed, relates to the speed and distribution of capital flows throughout the international business community aided by e-commerce, the second is linked to the inevitable expansion of services at specific major hubs, located strategically along international and transcontinental routes.

This chapter will attempt to review some of the operational implications of these changes, and their impact upon the increasingly complex roles of the major airports of East Asia. In doing so, the various topics under discussion will be treated as set of dimensional activities, often quite complex in terms of their designations as operational roles, but linked by their common relationship with the airport as the managing agency. The example of Seoul-Incheon airport, discussed in the previous chapter, offers a working model of such a customized and multi-operational agency, embedded as it is within its own mega region. In later discussion, Hong Kong will be used as a working example of various developmental strategies, given its already demonstrated role as a major mega hub in the East Asian region.

The Emergence of the Mega-urban Region

The phenomenon of increasing urbanization in East Asia has already been addressed in general terms as part of earlier discussions. It is characterized classically by the spatial concentration of FDI in the capital city and also the key international ports. It is regularly accompanied by the growth of core metropolitan regions, which stretch beyond conventional administrative boundaries, often into distant areas. These can stretch (Douglas, 1998) up to and beyond 80 to 100 km from the metropolitan core of the mega-urban centre. In fact, distances of up to 300 km are now becoming viable.

The PRD offers an excellent example of mega-urban growth. Located at a key point in the estuary, Hong Kong is an extended metropolitan area that melds into the major province of Guangdong. Multi-lane perimeter expressways and rail services run along the eastern arm of the delta and link the city to Guangzhou's metropolitan area. On the western side, a second arc of expressways makes a further important linkage with the ports of Zhuhai and Macao, a second strategic autonomous region. The strong sense of physical continuity is reinforced by the existence of multi-modal service linkages that allow a tourist or business person to leave any one of the cities involved and return to it by utilizing sea, land and rail or air services.

The increasing emphasis in other East Asian countries on the building of new inter-modal transportation systems has seen the further development of competing hub centres at strategic points where there is a convergence between regional and local transport networks. Operational multi-functionalism at a specific hub centre that is serving both international and domestic traffic allows for a seamless balance of activities. In other words and in the case of a major airport, it can carry out its primary modal function as an dedicated aviation centre, while maintaining direct operational linkages (Rodrique, 1996) with alternative road, rail or sea services.. In doing so, the airport is fully utilizing its two key locational advantages of centrality and immediacy.

Within such an urban region, a major hub such as Hong Kong can effectively transit air, sea and land and have an area of influence that covers several regional cities at a time. In doing so, airports can make strategic use of what has been called (Rodrigue, 1996), the existence of articulation points. These can best be described as an infrastructural interface between different spatial systems, acting as a gateway to each in turn. In addition, articulation points have a further important role as logistics identifiers for value added services.

The span of territorial influence, exerted by an international articulation point such as an airport, conventionally includes both a foreland, which contains its operational space and a hinterland, where usually its various traffic flows have a range of destinations and clients. In doing so, it fulfils its basic traditional and important role as a key generator of O-D transactions; without this, such vital industries as international tourism would simply be unable to function.

Defining Technological Progress as the Engine of Industrial Development and Economic Growth in East Asia

Some of the earlier chapters have given consideration to the geopolitical and other issues that have shaped the economic progress of the nation states that are located in the East Asia development corridor. Inevitably, the contextual focus was shaped by the mainstream literature, which saw the industrialization as an evolutionary process in which modernity and efficiency was really a function of technological catch-up. By inference, this would presume a willingness by venture capitalists and governments to plan a development programme moving the modernization process forward from conceptualization and the initiation of production, beginning with a required period of research and development (R&D).

The motivation as defined in the conventional international business textbooks would see FDI initiated by multinational corporations conventionally in search of lower unit labour costs and coupled with expectations of a new opportunity for market growth. In addition, national governments would be expected to offer financial and planning support as a quid pro quo, with appropriate taxation incentives built into any formal contractual arrangements.

A countervailing view (Mathews, 2001) suggests that, in fact, this approach was really bypassed in many of the East Asian countries. By contrast, the leading-edge members of the ASEAN group would use technology acquisition and the adoption and adaptation of technology according to their own strategic designs and specifications. The shift over time to knowledge-intensive industries would therefore involve a collective approach in which knowledge is spread and diffused across sectors. This is because material benefit is sought form the economy at large, which requires what Mathews calls a process of technical diffusion management (TDM).

This approach entails a significant shift in theoretical perspective; the focus moves away from the classical western model of the single firm and addresses the viability of operational clusters, each supported by their own requirements in terms of institutional support. The important distinction here, as already noted, lies in the location at strategically important mega-urban sites of a number of industrial firms. Each one would be following specific cycles of production activity, often within a very wide network of production, distribution and delivery to the final customer that in transit may actually cross the physical borders of a number of mega-urban centres, both nationally and internationally.

Earlier discussion has already commented on the basic modality of airport provision as the servicing of passenger needs either at the beginning or completion of a planned O-D sequence, or alternatively providing in-transit and between-hub support. There is now a need to expand that role into the second operational dimension, namely the servicing of producer needs, demand for which has seen aviation logistics and supply chain management become increasingly important on a global basis.

The Economic Activities that Develop in the Airport Area

It is possible in terms of its spatial geography to conceive of the airport as a set of activities that become more complex as a function of their various operational roles. They have also, as some of the earlier discussions have indicated, begun to expand in a number of directions that, in economic terms, are influenced directly by the constant pressures of real-time and customer proximity.

The influences of multi-modal transport availability add a mobility dimension to the role of the airport. These, it is suggested (Weisbrod et al., 1993), may be defined in the sequence shown in Figure 6.1.

The degree of business concentration in the various categories of activity relates directly to the form of the service or production function that is served by the geographical location. These are shaped dimensionally between services that directly support the operational dynamics of the airport and those that are located in order to maximize the direct time transit efficiencies of their high-value and often relatively low-volume output.

From the point of view of the air cargo/freight dimension, the need to address time-specific needs in the delivery of services is increasingly becoming a major strategic issue for all airports. Within this context, it has especial significance for the East Asian mega hub, since there is an important degree of competitive advantage involved. The linkage between good cargo/freight capacity (Kasarda and Green, 2004) and the ability to take advantage of the new fast-cycle logistics and supply chain is of vital importance.

1. Based at the airport.
   Relating to the total range of service functions for air traffic based on the volume of activity as well as regional considerations as a functional hub.
2. Services adjacent to the airport.
   Including aircraft catering, maintenance, passenger and employee services, airport-related freight and mail services.
3. Services within the vicinity of the airport.
   Located either in dedicated centres close to, or along, an access corridor and serving the growth of employment both at the airport and in ancillary businesses locating in the area.
4. Services elsewhere in the metropolitan region.
   Having their specific purposes actively shaped by both regional and airport market considerations, such as BtC transit services

Figure 6.1   The location of activities relative to the airport
Source: Weisbrod et al., 1993.

From a macroeconomic perspective, the role of any aggregate growth in a country's volume throughput and return from air cargo is directly correlated with the growth of national GDP. As a demonstration of this effect, and using Hong Kong as the working example, we find that between 1992 and 2003, allowing for the downturn in 1997–98, air cargo as a percentage of total trade rose from 17.7 per cent to 30.3 per cent.

Its identity as a global airport is reinforced both by its location on the north south axis of the major international and transcontinental scheduled air routes, and by its performance as an international leader in both passenger and air cargo services. Table 6.1 reflects this status.

Table 6.1   The economic performance of Asian cargo airports in 2004

World rank	Airport	Tonnage	%Change
2	Hong Kong	3,066,734	15.3
3	Narita	2,355,398	10.5
5	Incheon	2,104,254	16.8
8	Changi	1,778,386	9.2
11	Taipei	1,702,405	17.5
15	Pudong	1,568,396	39.3
19	Don Muang	1,039,351	10.1
22	Osaka	874,995	11.4
24	Haneda	769.084	8.0
27	Beijing	685,741	7.9

Source: Air Councils International, March 2005.

The significance of the world ranking segment of the table lies in the fact that while the mature East Asian mega hubs and the Korean winged city are ranked 2, 3, 5 and 8, the major Chinese airports at Shanghai and Beijing are ranked at 15 and 27. In turn, it is clear that the latter major hopefuls for future international cross-border status (Air Councils International, 2005), Baiyun International and Shenzhen, have to modify their earlier expectations of an early entry into world class status.

In the Baiyun case, which opened in July 2004, there is a need to look to the medium term for the attainment of that level of traffic growth that is a major indicator of international recognition. By contrast, the future of Shenzhen is materially affected by its changing emphasis away from textile and clothing production and towards high-technology R&D. It is also materially affected by its relatively short distance (37 km) from Chek Lap Kok, and the possible integration of both sites over time as the estuary of the PRD develops.

That competitive pressures to become regional and international powers in the world's international air freight markets are increasing the number of new entrants for the status of megahubs is reflected (ACI-Europe, 2004) in the short- to medium-term trends in global traffic. Projections for unconstrained growth (ACI-Europe, 2004, p. 11) in the period 2005-2007, indicate a mild slowdown, from 9.5 per cent to 6.6 per cent.

But the medium-term projection (ACI-Europe, 2004, p. 12) sees a movement downward from 7.5 per cent to 7.0 per cent in the period 2007–10, and a consistent average of 5.4 per cent for the period 2010–20. What this means in terms of the increase in traffic volume is an annual growth in freight over the next 15 years, from 75 million, to 174 million metric tonnes. An immediate word of caution must be inserted into the discussion at this point. It is acknowledged that unconstrained projections do not accommodate an imperfect market. In other words, there is an implicit assumption that, as demand grows, airports and facilities will grow in tandem to meet operational demands. These include a range of regulatory, political, environmental and geographic issues that have to be carefully considered.

In terms of passenger growth, for example, constrained capacity shortfall on an international scale ($N = 273$ airports) could exceed some one billion travellers by 2020. In turn, freight volume could be restrained to 3.8 per cent per annum from 4.6 per cent, with an attendant effect on movements (2.2 per cent not 3.0 per cent). Discussion will return to the policy implications of these projections later in the chapter.

Modelling the Infrastructural Changes needed by Air Logistics: The Possible Emergence of the Global Transpark

It is timely to recall that while Hong Kong retains a somewhat unique identity as a major centre for international trade, business and transportation, which has been reinforced by its status as a Special Autonomous Region, it remains essentially in that geopolitical condition as an important part of China. In turn, while China is contiguous with East Asia, it shares spatial locations and borders, with a large number of other states and regions. From an economic perspective, it also ranks as an emergent superpower, conterminous with Japan and India.

It is noticeable, in the wake of the 1997 transfer of power and control over Hong Kong back to China, that a degree ambiguity exists over the fact that China now has two major international airports of mega-hub size within the PRD. The presence of Baiyun International, which opened in 2004, has been accompanied by strategic statements about its growth potential that seem to threaten the pre-eminence of Chek Lap Kok. This is reflected in the considerable interest of the major freight carriers, as well as US airlines, with regard to the new hub's potential, although to date, the expected rush of new flights and services has not eventuated. On the other hand, the fact that Baiyun has the potential to expand, and indeed is expected in the future to have four or even five runways, is reflective of the key role of the PRD as one of the most important economic regions in the world.

The question immediately arises is, what then is the future of Hong Kong? One proposal that attempts to deliver the answer suggests that it might lie in the maximization of the extended metropolitan region (EMR) as a totally integrated transportation system or global transpark.

The Global Transpark as an Operational Means of Air Cargo Growth

The concept of a global transpark (GTP) was first proposed in 1992 and a number of prototypes were built. Of these, Subic Bay in the Philippines is an important model, because as a key base for FedEx, it is reflective of the way in which the major air freight MNEs, are establishing their own mega hubs (Sit, 2004) at the global interstices of their business (see Figure 6.2).

1. A multi-modal transport hub with air cargo facilities at the core, linked directly to efficient expressways, rail and water transport, integrated as a new transhipment centre.
2. An advanced telecommunication and e-commerce centre, linked to the hinterland and to the world, for instant data assembly, processing and transmission with a view to extending and intensifying market coverage and material sourcing.
3. Development of industrial sites for agile manufacturing for a global market.
4. Development of a trade and warehouse centre for both the regional and the global market.

Figure 6.2   The four infrastructures of the GTP
Source: Sit, 2004.

The functional design of the GTP would be seeking a seamless interface of all transport modes, with further support from state-of-the-art customs and other clearance procedures. In addition, seventh freedom rights would be accompanied by co-terminal and ground handling rights. In addition (Kasarda, 1998), the existence of in-transit bonded status for transhipments would be locked both on and off site where in-transit industrial processing is required.

In fact, the concept advances beyond the traditional notion of an air cargo facility and evolves as a crucial logistics and manufacturing plus retailing facility, thus becoming a new form of network hub in its own right. The argument that China needs such a development and the PRD is the most logical place to develop such a site is based on the fact that, to date, the development of a significant air freight sector in the PRC aviation industry lags behind the rest of the region.

This raises immediate problems for the existing facility at Hong Kong in the physical sense, since it is located on a customized site – an artificial island that cannot be expanded without enormous cost. On the other hand, the extended metropolitan region (EMR) is situated at a global confluence of air routes linking Europe, the United States, East Asia and Australasia. It is also a matter of some considerable comparative advantage that under the Basic Law of 1997, which now defines its status, it has retained its independent air rights.

The further existence of the group of airports now commonly identified as the A5 allows a degree of choice for site. A development programme in a location such as the coastal airport at Zhuhai, on the western side of the estuary (Sit, 2004), for example, would allow for the following key roles and functions to be developed.

The need for detailed coordination covering the A5 and related cities is clearly identified. At the same time, there is a further need for the PRC central government to provide at minimum a set of conditions for the operation of the GTP. These would include the granting of flights and slots for the international air cargo integrators such as FedEx, UPS, TNT, and DHL in and out of Zhuhai, as well as the rescheduling of Zhuhai as a priority cargo hub with a subsequent diminution of passenger and domestic cargo business.

1. To become Asia's air express hub transhipping air express cargo to the bulk of Asian cities, which are within 3 hours O-D flying time.
2. To develop as Asia's major logistics hub for MNEs servicing the Asian markets.
3. To evolve as Asia's major e-commerce base.
4. To become Asia's JIT assembly location for high-technology–high-value products.

Figure 6.3   The future strategic functions of the GTP: The Zhuhai model
Source: Sit, 2004, p. 159.

The establishment of a free trade zone as well as bonded status for some activities would need to apply to all cargo activities related to international traffic. It would also need to be extended further to bonding arrangements for a fast speedboat service between Zhuhai, Macau and Hong Kong airports.

The GTP model seems to fit within the larger discussions that are going on in the PRD with regard to the future integration of the region as a super-zone. It also gains a degree of traction from the fact that a new civil aviation agreement was signed in late October 2004 between Hong Kong and China, which formally recognized the status of Hong Kong, as a key Asian hub. The agreement was signalled first by the award of 12 weekly freight slots at Beijing International to Cathay-Pacific. This was followed by an equity arrangement between Air China and Cathay, which suggested (Thomas, 2004) that both airports were heading for dominant hub status in Asia's largest growing freight market.

De-verticalization, Fragmentation and the International Division of Labour

The previous discussion has covered some of the organizational possibilities that are being driven by the development of what is popularly called the 'new' or 'knowledge economy'. While classical economic theory assumed a market where information was both costless and known by all participants, the reality revealed the limitations imposed by symmetric information in markets where firms in fact operated at arms length from each other when it came to information sharing. By contrast, an efficient and, by definition, optimal competitive market economy in a world of e-commerce, real-time access and the Internet depends upon the availability of information to all economic agents at optimum times of need.

The operational design cited in the GTP model also potentially allows for an increase in net efficiencies, lower transaction costs and increasing returns to scale. The reduction of transaction costs alone allows operators to more than offset (Lau, 2003) the increased transportation and communication costs of servicing demand on a global basis. In addition, the existence of information and its increased availability allows for two-way integration between tangible capital, such as developmental capital and intangible capital in the form of human, R&D and knowledge capital.

In the case of human capital, its presence or absence can be critical for any successful new development, such as the building of a new airport, that is intended to maximize to full advantage state-of-the-art information technologies. The creation of a human resource network to turn planning into physical reality is by definition a form of human capital, which will not only create an airport, but also a series of network externalities. This is because the much higher skill requirements of the assembled workforce will entail over time the further creation of a critical mass of knowledge over and above that required by day-to-day activities. It is also important to recognize the fact that today's well-qualified labour force has present in each of its members a depository of individual human capital, which can be transferred at will to any place in the world where specific skills aptitudes and industrial experience are currently required. At the same time, this mobility factor, which has been made possible by major advances in aviation technology, strikes a warning note. This is particularly true for those member states in East Asia, as elsewhere in the world, which do not recognize the fact that today's high skills demand a global premium and with it a high degree of competition for available talent.

The Globalization Effects of De-verticalization on the Production Process

The traditional textbook firm is a vertically integrated organization in the sense that it does not go outside its own organizational perimeter to purchase intermediate goods or services. The classical example is General Motors in the 1930s and after. General Motors comprised a set of subsidiaries that manufactured all of the components of the its automobile range up to final assembly and distribution through tied dealers. Its one link with outside firms would be in the market for raw materials such as metal ores and rubber.

By contrast, de-verticalized firms purchase an increasing amount of their requirements for intermediate goods through outsourcing its orders to other firms.

This restructures the division of labour by a major shift away from the vertical division within a hierarchical organization to a horizontal division, where work in design, manufacturing, marketing, inventory transportation and distribution functions (generalized outsourcing) occurs both within national and across international boundaries.

As a consequence, logistics and supply chain management involves the control of production processes that may be distributed across a range of firms at any point in time. This raises the question of the need for operational standards that will require the need to allow for precise and accurate performance rules, with benchmarks set, preferably, by independent agencies employing systems such as ISO.

There can be no doubt that de-verticalization impacts on all forms of business activity, as the centre of organizational gravity in the region's aviation industry continues to re-shape itself from commodity based to value-added client demand. Airports face the same patterns of change as all other service sectors in the aviation industry, and must deal with these within the parameters set for their complex of roles.

They must also, in keeping with other industries, face the undeniable fact that the need to reach quality standards is open-ended. In other words, there is really no point in future time where management can switch to cruise mode. Quality standards require persistent maintenance and sometimes modification as client perceptions of need change over time. How far this important lesson is understood by airports would be a timely and useful question for future research and evaluation.

Some Problems with the Stage of Development Phase of Air Transport Infrastructures in the East Asian Development Corridor

It is clear from discussion in earlier chapters that the various states that share the East Asian development corridor are at various historical stages in their development. As a consequence, there are significant variations in the standards and quality of logistics both within and between countries. Where costs are high, the problems can be traced to (Carruthers et al., 2004) poor transport infrastructure, underdeveloped transport and logistics services and slow and costly bureaucratic procedures for dealing with the movement of both export and import goods. A recent report by the World Bank suggests that the various states can be characterized in the following group order, according to two important characteristics: outward-orientation and degree of accessibility. The pattern shown in Figure 6.4 emerges when the member states are compared.

1. Outward-orientation with high accessibility: Singapore, Hong Kong, Korea, Taiwan.
2. Outward-orientation with some accessibility: China, Indonesia, Malaysia, the Philippines, Thailand.
3. Limited accessibility: Vietnam,Cambodia, Laos PDR.

Figure 6.4   The pattern of orientation and accessibility
Source: Carruthers et al., 2004.

The pattern correlates such factors as per capita incomes and the commodity structure of trade with the key distinction to be found between high technology exports in 1, and limited resource-based commodities in 3.

From an airport perspective, the degree of access as measured for airports with paved runways in excess of 1523 m in length adds a further dimension to the pattern defined above. Quite clearly, Korea's plan to dominate the northwest looks feasible on the grounds that the number of paved runways is a good proxy for a positive governmental approach to airport development.

On the other hand, that assumption's validity is tested by the fact that Vietnam, for example, has developmental plans in place for a number of new sites in addition to its key hubs at Noi' Bai, which serves Ha Noi, and Tan Son Nhat, the metropolitan airport for Ho Chi Min City. These include new airports at Haiphong, the major port in the Red River delta, the former US military base in Da Nang and a further development to be located in the Mekong delta. Clearly, these projects will be dependent on key inputs from FDI and related sources.

By contrast, Cambodia and Laos PDR will require a very considerable infusion of capital, which in the former case may be somewhat limited due to a significant and continuing period of political instability. In turn, the depredations left behind by US bombing during the Vietnam war, as well as the long departed Pol Pot regime, were both widespread and intensive. In the latter case, the characteristic dependence on commodity outputs does tend to limit the available resources that might be put to use in order to develop a viable civil aviation industry.

Some degree of confirmation for these views, which are partially the result of personal observation while working as a consultant in the region, may be found in the ranking of viable runway facilities given in Table 6.2.

Table 6.2   Availability of paved runways at East Asian airports

Country	Rank	Number of runways (minimum length 1523 m)
Korea	1	375.7
Philippines	2	113.3
Thailand	3	79.5
Malaysia	4	63.7
Vietnam	5	45.5
Indonesia	6	32.8
China	7	27.4
Laos PDR	8	25.3
Cambodia	9	22.1

Note: Singapore and Hong Kong are not included, since they rank as major global locations recently awarded first and second places in an international airport rating exercise.
Source: Carruthers et al., 2004, p. 120.

The Strategic Importance of Air Freight

It is generally acknowledged that air freight is of vital importance to East Asia. This is because, while it is only 1 per cent of the region's international trade by volume, it exceeds 35 per cent by value. The strategic reason for its importance stems from the distance to key markets such as the United States and Europe. In addition, given the high proportion of manufactured exports requiring rapid delivery, their high value-to-weight ratio allows the ad valorum cost of air transport to be relatively low.

Given the importance of FDI in the form of high-value industries wishing to locate in countries with good air freight facilities, competition between airports in the region for the role of hub for major dedicated logistics companies is growing, with some of the smaller and recently constructed players experiencing faster growth than larger, but less-efficient, players.

In the context of opportunities for future growth, the major airports are clearly and increasingly focused on China as a key player. In a very real sense, China plays two important geographical roles. The first is that of an independent country moving toward the status of an economic and geopolitical super power in Asia. The second is that of a major force in the East Asian region because of the contiguous location of its primary growth regions in the southern and coastal provinces, which stretch between the PRD and the Shanghai-Yangtze delta. From a strategic point of view, the deregulation of provincial Chinese airports, with control transferred from central to local government, opens up the prospect of developmental capital being sourced from off shore through joint venture arrangements.

This pattern of development is reflected throughout the Asia-Pacific region, with market liberalization and airline expansion both under way. Capital expenditure on airport development in 2004 rose to US$31 billion, which is the highest total recorded by ACI since it began to collect data in 1995. Meanwhile, in marked contrast to the United States and Europe, the Asia-Pacific and Middle Eastern airlines returned an aggregate net profit of US$3 billion. This reflected a growth of 20 per cent in passenger traffic over the 2003 figures.

While many of the market signals look positive across most of the operational sectors of the aviation industry in the general region, with major growth being led by East and North Asia, there remains the important problem of fuel costs that continue to rise. The rising price has been offset to some extent by rising passenger numbers and surcharges on the price of travel. It is clear with hindsight that IATA's forecast of an industry wide loss of US$6 billion, set in May 2004, when the price per barrel had reached $US47, is well below the actual cost which is close to double that figure. The sometimes rapid pattern of price escalation since then has raised the average to US$53 billion, with the final number still unknown. According to the Secretary General of IATA, the effect of a US$1 rise adds an extra billion to airline costs. There will be further reference to and discussion of the problems associated with rising fuel costs in a later chapter.

A second important effect yet to be measured is the growing impact of LCCs, which are now entering the domestic market in larger numbers. The new start-up trend in the region has already signalled the birth of 26 new airlines in 2005. It is anticipated that the numbers will increase rapidly, as investors leave the saturated European market in search of expected and much higher returns in the Asian markets. There can be no doubt that East Asia will be a major target in the future, as 14 of the new airlines will be located in China, Hong Kong and Macau.

While fear of market competition seems to be based on the experiences of Europe and North America., it is suggested that a more primary concern will relate to long haul routes, which are unlikely to be affected by LCCs for some time. While premium yields are consistently strong, much of the yield growth has come from the appreciation of regional currencies against the US dollar. It is further expected that the pressure on premium yields will remain high; the question then becomes, are they likely to be diluted in the future by greater competition in the premium market and possible efforts of a 'Virgin' entry with the intention of offering lower-cost corporate travel?

The Intensification of Demand for Human Resource Skills as Regional Airports Grow

The expected expansion of airline fleets, which has been signalled by the major regional carriers, both AAPA members as well as non members, is expected to see 916 new aircraft enter airline service in the period 2005–09. Excluding fleet replacements, some 47.5 per cent of the new orders will see an additional 121 aircraft come into service every year until 2009. Some 88 per cent will be narrow-bodied and medium-sized wide bodies, with balance filled by ultra long-haul and very large types such as the A380 and the Boeing 747–400. Seat capacity will increase as a result, by 45.4 per cent, with the greatest demand found in the expanding number of high-growth markets

The impact on the demand for skilled labour has all of the symptoms of an impending crisis. This is because market competition demands higher levels of productivity from all aspects of airline operations. This places intense pressure on the human resources of the airlines at all levels of skills and raises the further possibility of constraints on their ability to respond to the level of demand (see Figure 6.5).

Additional pilots	10,000
Additional cabin crew	21,000
Additional maintenance crew	15,700
Total new employees overall	94.200

Figure 6.5   Estimates of employment vacancies in Asia-Pacific, 2005–09
Note: The highest level of demand is found in North Asia, including China and Japan, followed by India and the Middle East and then Southeast Asia. This pattern clearly follows the current distribution of traffic growth for both passengers and freight.
Source: Centre for Asia-Pacific Aviation, Sydney, 2005.

The possible consequences that might arise from this general and pending crisis include a range of slow downs and service delays, with important development projects either postponed or simply abandoned. In addition, the effective rise in demand, coupled with supply-side failure, will exert an upward pressure on labour costs, with an attendant rise in wage and salary demands. Further policy shifts might require the need for training programme acceleration, with third-party suppliers engaged to plan and drive such activities.

In addition, the outsourcing of operations to foreign and/or contract labour might require, in turn, that airlines relax any regulatory restrictions imposed by government, or policy restrictions by the company on the employment of foreign workers. The onus for all of the decisions relating to human resource needs clearly falls on the individual airlines, but it would be foolish in the extreme for airports, not to respond in an appropriate way to the problems faced by their key clients.

Discussion has ranged very widely across the various and complex issues facing airports in today's highly competitive regional market. There is now a real need to turn the lens onto several new and related questions in the next chapter. The direction of discussion will refer, for example, to the important need for the parties involved in the development strategies of major airports in the region to access levels of future investment that will guarantee the ability to respond to the demands of further projects. The complex relationships between such resources and the political influences that often shape decisions will be further examined. Finally, consideration will also be given to the large question of the future role of private as opposed to public ownership of the increasingly multifunctional enterprise that constitutes the modern international airport in East Asia.

Chapter 7

Political and Market Issues Confronting
East Asian International Airports

Introduction

When the important questions surrounding market deregulation become the focus of policy discussion, the ends–means issue that most frequently comes to the surface always involves the degree and extent of private investment in infrastructural projects. This is particularly reflected in the various strategies proposed for airport development during the last quarter of the twentieth century. Throughout that period, there was a continuing debate as to the important role that private capital should play in the construction and expansion of national airport systems.

According to the World Bank (Silva, 1999), the 1990s was a period which saw private sponsors participate in projects involving 89 airports in 23 developing countries, worldwide. The aggregate value was US$5.4 billion, with 45 per cent going to the Latin American and Caribbean regions, 22.5 per cent to East Asia and the Pacific and 21.2 per cent to Europe and Central America. Such developments signal a growing governmental interest both in FDI and in other forms of subsidized and expert assistance through formal interventions and targeted projects. At the same time, it is important to bear in mind that private participation in the infrastructural development (PPI) of the airport sector, as is the case in the other service industries of developing countries, is still in quite early stages of its redevelopment.

Unfortunately, the data reported above does little to identify either the real dimensions of true investment need, especially in key industrial sectors, or the range and complexity of the issues confronting the members states in each of these regions. These often have a significant influence on the planned outcomes of FDI, which may in reality end in a serious gap between a major development project's goals and its final outcomes. Such possibilities have a serious significance for FDI, especially where the capital cost of a new airport is concerned.

In this chapter, an attempt will be made to identify, albeit somewhat briefly, the problems arising through the uneven patterns of infrastructural development that are found at the level of individual states. They affect most notably those countries whose economic progress has been constrained by historical, geopolitical and limited natural or spatial endowments. Their effective resolution is vital in the geopolitical sense, since they underpin and signal the need for appropriate sectoral and industrial strategies to alleviate their problems. The decision to do so immediately raises the major question of cost and the potential role in the absence of effective capital markets for FDI.

Initial discussion will look at FDI requirements in the infrastructural context and with a strong emphasis upon the need for a best-practice approach to planning. This will then lead into the important question of market liberalization, as a natural corollary to FDI, with the primary focus on the evolving roles of the EU and the United States as leading promoters of deregulation in the cause of more international flexibility in aviation markets.

This will be followed by a further consideration of both the effects of such changes, especially the US promotion of bilateral open skies agreements aimed at the national states in East Asia. The context of discussion then is expanded to consider the pivotal position occupied by China, as a state, with a massive southern coast contiguous to the region, joined to an emerging role and identity in the global sense as an economic and geopolitical superpower.

The balance of the chapter will finally review the strategic role of the major airports of the region as potential mega hubs, required to service the requirements of both national and cross-border hub-and-spoke clients.

Current Best Practice Problems with Policy Planning and Coordination in East Asia

The pace of economic development in the ASEAN community, and especially in the East Asian corridor, has clearly picked up since the financial crises of the late 1990s. At the same time, a fundamental paradox has emerged. This relates directly to the extent to which any national infrastructure is able to sustain the increasing demands made upon it by economic growth. Given the structural shifts from commodity-based to value-added export production being sought, and often with considerable success, by the majority of states, especially the ASEAN-5 members, key sectors such as transport and, within it, aviation are under considerable and growing stress, as a function of increased demand.

In addition, the problems related to policy planning and coordination, are appearing to indicate either emergent or current 'deficits' (Asanjuma, 2004) in infrastructural provision. These are signalled by a number of backlogs in delivery of services, leading to under-supply and scarcity on the one hand and over-supply in some cases and wastage on the other. The problem is compounded by the fact that such limitations are not evenly distributed across sectors, even within individual countries. Both China and Thailand, for example, the latter country the former epicentre of the fiscal melt down that commenced in 1997, have engaged in massive investment in infrastructure over the last 5 or so years. By contrast, both the Philippines and Indonesia have been less successful. To what extent geopolitical uncertainties have influenced the situation requires a much more detailed analysis than is possible here.

Some Proposals for Cross-sectoral Private Sector Participation

The need for a best-practice model to overcome the limitations discussed above has been on the development agenda of most of the major international agencies. This has given raise to a Private Sector Practice (PSP) model aimed at a division of responsibilities in which national governments look after requirements for reconstruction planning, efficient monitoring and regulative procedures. The resultant PSP requirements are then carried by appropriate and carefully selected private sector agents. A recent study (O'Sullivan, 2000) highlighted the importance of effective regulation joined to competitive tendering and operational transparency. It noted the key elements listed in Figure 7.1.

1. Create private sector management, investment, construction and financing.
2. Ensure transfer of responsibilities through open competition.
3. Create management contracts, capital leases, concessions, sale of assets and rights to operate.
4. Transfer commercial risks to the private sector, with any other risks assigned to the best party able to mitigate them.
5. Begin development of long-term financing sources.

Figure 7.1   Benchmark requirements for effective PSP controls over FDI
Source: *O'Sullivan, 2000.*

By contrast with the other transport sectors, aviation presents a more dynamic picture in terms of infrastructural development, as was noted in the introduction to the chapter. But, as the following discussion will reveal, there are specific legal and administrative rules of compliance, especially in the airline industry, that are reflective of the major forces that impose specific degrees of constraint on the operating ability of national carriers. In fact, they impede the ability of some airlines to maximize comparative advantages in an increasingly international market.

It follows that the infrastructural problems facing the aviation industry are not imposed to any great degree by spatial, human resource or natural endowment factors. Instead, they are shaped by geopolitically designed and imposed rules, set prior to the development of modern national aviation industries in the region, but now requiring compliance under the authority of supra-national agencies. These in turn are authorized to award licences to provide services, subject to audited controls by duly empowered national civil aviation authorities (CAAs).

This, of course, is not to argue that the removal of such regulative restraints would naturally contribute to efficient and effective industrial growth. The fact may be somewhat obvious, but it does have to be remembered that air travel is not totally devoid of risk, either physical or, to some degree, mental. The establishment of universal standards of operational and environmental safety has long been a contributor to the general level of efficiency exhibited by airline and airports alike.

Perhaps the more important question, which will be discussed later, involves the extent to which operational standards can be relaxed in order to test the degree to which their modification will improve operational performance.

East Asia and the Current Regulatory Problems Limiting FDI in Aviation Services

The element of paradox deepens when we approach the question of a financial strategy supportive of the aviation industry in East Asia. This is because the international industry sees itself caught up in an operational duality. The global nature of aviation is increasingly reflected amongst major airlines, through their membership in strategic alliances that involve multiple and complex cross-border arrangements, such as code sharing and joint marketing agreements. These are driven by rules that restrict forms of equity transfers between airlines. As a consequence, the conventional process of competitive growth through mergers and acquisitions has been notably absent.

The source of these limitations can be found in the fact, (Findlay and Goldstein, 2004) that they are bound by a complex web of regulation, both national and international, which has offered over time quite considerable resistance to any forms of FDI. The reasons are found in traditional concept of air transport as a form of public utility, with a long tradition of statutory control.

In the American case, federal regulation literally shaped and controlled the airline industry until 1978 and the deregulation of the market following the passing of the Airline Deregulation Act. Economies of scale were also an historical casualty of a long tradition of regulative control over the size of corporations, which was based, in turn, on a fear of the effects of monopolistic competition in the market.

There is a consequent danger, as British Airways will attest, in trying to create a merger with an American legacy carrier, that a deputed federal agency will bring a counter-claim on the grounds that such a merger would be in contravention of national anti-trust legislation. Access by foreign investors to shares in the major carriers is also limited in law to a permanent minority holding. Given the fact that the United States has been the major leader in the push for global market liberalization since 1978, its adoption of open skies bilateralism as the main strategy for advancing deregulative arrangements has tended to delimit the terms of reference in the agreement to the immediate parties.

This has also led to a strong and persistent emphasis upon the limitation of the applicable conditions introduced by the agreement exclusively to the national signatories. It should be noted, at this point, that these rules are inevitably applied without distinction in the case of ASEAN-5 countries, which are the homes of several highly successful international airlines, three of which are members of international alliances. By contrast, the flag carriers of the less-developed states have often been found to be at some disadvantage with regard to market access beyond the gateway hubs, both under the terms of the agreements and through limitations in appropriate fleet sizes.

There are now indications as further discussion below will indicate, that the American concept of air service agreements are being increasingly broadened to include complementary terms relating to trade and other forms of market access. Needless to say, both the ASEAN-5 group of countries and China-PRC have become favoured nations, given the fact that anticipated traffic growth in the medium to long term is the highest in the world. They also enjoy a degree of comparative advantage as major locations for the establishment of production centres by a whole range of international corporations seeking a competitive return from expected lower labour and production costs. This pattern is now expanding to include, notably in China, retail chains and fast food outlets.

We can now return to the question raised earlier, what then are the needed modifications to regulations that might be changed with a positive effect in terms of operational efficiencies? These are as listed in Figure 7.2.

1. Dominant state ownership (outside the United States) with regulative and often symbolic protection where the government owns a national flag carrier.
2. Controls on entry, capacity and tariffs, imposed by investment needs, and the effects of network externalities.
3. The existence of specific international regimes such as IATA, ACI, ICAO and numerous regional organizations.
4. A significant pattern of vertical integration within the ownership, management and the physical infrastructure of air and ground services.

Figure 7.2   The normative conditions limiting FDI in civil aviation
Source: Findlay and Goldstein, 2004.

The Powers of the Bilateral Air Service Agreement

The regulative process governing air services in the international route sector remains the Bilateral Air Service Agreement (BASA). By mutual arrangement between governments, the national airlines as parties are nominated to operate on a given route, with terms of reference covering capacity and frequency. It also covers (Findlay and Goldstein, 2004, p. 41) the setting of capacity and the charging of fares. Further restrictions apply to third country airlines with regard to their access to route. The BASA also covers such matters as royalties, traffic compensation, traffic and revenue pooling. Increasing capacity, in turn, must be subject to mutual agreement by the parties.

The restriction on foreign ownership is typically demonstrated in the requirement that any BASA be between designated carriers that are substantially owned and controlled by the nationals of the countries involved. The alternative course of seeking entry into regional as opposed to major alliances that will, to some extent, reduce the revenue problems imposed by BASAs, carries a degree of risk, given the fact that outside the major groupings such as Star, One World and Sky Team, the history of such arrangements tends to be fraught with instability and consequential uncertainty.

It is also a matter of some interest that out of the 16 largest airlines flying from their national hubs in the East Asian region, only three, Singapore, Cathay Pacific and Thai International, are members of a major alliance. There are some indications, however, that such membership might be a good idea in an increasingly competitive market, and is being taken up by airlines such as Air China.

Considerable pressures have been growing over time to modify the terms under which the airline industry has operated, in favour of more open skies arrangements. When used in the context of a notional BASA, the term is in fact somewhat ambiguous. This is because it is an essentially elastic concept that extends to include cabotage and goes significantly beyond the fifth freedoms that contemporary bilateral agreements employ. As a final comment, perhaps the most delimiting effects of the legal inability to enter the international foreign capital markets may be found in the current state of the cash-starved US legacy carriers. It finds poignant expression (Havel, 2003) in their last ditch attempts to sustain themselves by defaulting on staff pensions and reducing in-house employment, in between frequent trips into Chapter 11 bankruptcy.

Legal and Political Initiatives to Institute Major Changes in the Current Operational Practices of International Airlines

One of the most potent forces that has emerged more fully in the twenty-first century aviation industry has been the deeper integration (Dymond and de Mestral, 2003) affected by the major international airline alliances. The creation of a worldwide system based on hub-and-spoke networks allows for the coordination of flight schedules around a few key hubs, which act as central locations. This creates problems for airlines outside the alliance networks, since they remain limited in their operational scope to point-to-point services. It is now estimated that more than two-thirds of international traffic (Dymond and de Mestral, 2003, p. 10) is carried by the alliance members of four such major organizations.

Perhaps the most profound proposal for the virtual abandonment of the BASA system has emerged from the European Commission, and the European Court of Justice. Previous comment noted that Europe has had a cabotage-based open skies regime since 1997. More recently, on 5 November 2002, the development of fifteen BASAs between the United States and individual member countries of the EU, which began in 1992, was declared to be in direct violation of some aspects of both European Commission primary treaty law and other secondary legislation. The impact of this finding on some 1,500 other such agreements was also noted.

The European Court also made two further unequivocal decisions with very wide implications. It stated that national ownership and control matters in the agreements under review violated a central principle with regard to the freedom of establishment of corporations. The act of designating as parties only airlines subject to the ownership and control of signatory states or its nationals effectively denied the rights of the other 14 states to receive national treatment at their hands.

An additional and very important ruling was attached to the judgement, which, in effect, allowed the projection externally beyond the purely internal jurisdiction of the Court. What is left open to interpretation and further argument is the extent to which the Court's ruling covered aviation matters. On the other hand, the Court does have a strong role in deciding the future of external air transport services.   What was a landmark decision then led to further action by the European Commission, in which a mandate was sought to strike out existing BASAs between the EU and other countries and seek to replace them with EU-wide agreements that would be designed to liberalize both the transatlantic market and to liberalize the rules relating to investment. Quite clearly, the expectation in Europe was to see market liberalization advanced in keeping with its philosophy of a single integrated market.

The EU Reform Package for the Airline Industry

After due consideration, the Transport Ministers of the EU member states agreed on 5 June 2003 to a three-part package of proposals for change requested by the European Commission. The first major agenda called for the opening of negotiations with the United States, with the aim of creating an open market. This would permit American and European airlines to engage in cabotage arrangements with regard to the service any pair of airports in the Unites States or the EU.

Under the terms of the second mandate, the investigation of the nationality requirements contained in some BASAs was called for, with specific reference to the countries involved. The third called for the Commission to give regulative effect to a rule that would allow bilateral agreements to continue, but under EU coordination and on the basis of a standard textual format.

In the event, the United States expressed its readiness to engage in the development of a new approach (Slane, 2003) to the question of the management of air services, and this was confirmed with a later statement by President Bush and the presidents of both the European Commission and Council, which confirmed that negotiations would begin in late 2003. The degree and extent of American commitment remains somewhat ambiguous, especially in the crucial matter of how far the market liberalization process might be allowed to proceed under what is now becoming an extended form of negotiation, which has now moved sequentially into meeting schedules for 2005.

The Prospects for the Unified Advancement of Aviation Market Liberalization in East Asia

It is an often repeated irony about the airline industry that while its absolute indispensability as a major force in the globalization of world trade and business is generally recognized and applauded, the industry as a professional entity has been somewhat reluctant to fit its own strategies to that dominant role. Within the context of the East and Southeast Asian region, as in other areas of the developing world, both airlines and airports have an enormous symbolism and political identity as icons for national pride. Serious divergences exist in the fact that while attempts to expand bilateral agreements with foreign countries seem to be popular, closer ties with neighbouring states have not produced strong initiatives. A useful example of this reluctance can be found in the case of the MALIAT agreement with the United States. The only recent event of note has been the departure from membership of Peru.

Quite clearly, the absence of a supra-national professional body linking the region's nation states with ASEAN-5 countries at the core severely limits progress. But, as discussion in an earlier chapter has already observed, the process of unification is not really helped with it 2003 declaration of global principles for the liberalization of air transport. In short, the advancement of needed changes are defined as the prerogative of each state, which has the right to choose the pace of such progress at bilateral, regional pluralistic or global levels 'according to circumstances'.

As a consequence, while the United States and the EU clearly advance their quite distinct strategies for the freeing up of competitive aviation markets, it would seem that the majority of states in the East Asian region will remain in what will become an increasingly reactive set of responses as activities generated outside the region continue to shape their aviation futures.

There is some need, for example, for individual countries to widen the terms of their responses to the continuing expansion of bilateral open skies agreements with the United States. This is because such agreements appear to be strategically viewed as the industrial means to larger geopolitical ends. The ultimate goal that has been confirmed officially (Larsen, 2000; Mineta, 2005) appears to involve the development of a transportation infrastructure, which can then become one of the major instruments that will bring about a region-wide economic integration by 2010 or 2020.

This is reflected in the increasingly parallel development of agenda items that include a range of commercial as well aviation matters. As an example, the United States completed an open skies agreement with Thailand on 9 September 2005. In the advisory statement (US Department of State, p. 1), it was announced that the parties expected the promotion of trade, investment, tourism and cultural exchanges to flow from the agreement.

An important example of this trend can be found in the proposals for what the Bush Administration has recently called a fully liberalized air service agreement with China. There is already in existence a BASA signed in 2004, which promised that it would be fully operational by 2011. The following decisions have since been implemented as a part of that progression. There has been an increase in the number of airlines permitted to service US–China routes from four to nine, with weekly flights to be increased from 54 to 249. Cargo hubs have been established in China by the major US carriers. Unlimited code sharing and the right to fly to all destinations in the signatory countries have been established. Further concessions are available to be granted to US carriers willing to operate in the less-developed regions of western and northeastern China. Finally, the estimated return to US airline carriers is of the order of US\$12 billion over the period to 2011.

Meanwhile, the view as of 2005 in Washington is that the proposed negotiations planned for 2006 should look at increasing the scope of liberalization to permit unlimited entry for carriers and services in response to market needs. At the same time, it would appear that progress toward full liberalization is still somewhat constrained in pace and tempo. This somewhat contradictory approach would seem to flow on from the US government's willingness to be directed by the appropriateness of individual and national perceptions of the best means to attain a more 'liberated' airline industry.

A second important issue involves the large question of the ability of national carriers to respond as the regional market expands and grows. As a matter of necessity, they will need to develop a strategic response towards increasing competition not only from airline strategic alliances, but also from other major international competitors all seeking to expand on the degree of comparative advantage already obtained through their existing BASAs.

Many of these issues are fraught with ambiguities and geopolitical factors that do not allow for the development of future predictions that can be endorsed with any degree of real certainty. They have very significant implications for airports in the region for the following reasons. The threat of civil disorder and terrorism is very real in a number of ASEAN member states. It has been compounded by natural disasters and by innate political instability in a number of cases. The cost of increasing airport security, coupled with the rise in the cost of fuel are other important issues that need to be factored into any future assessment of competitive status and the medium- to long-term consistency of high levels of demand for services.

Thus far in discussion, attention has been directed toward the East Asian region as a collective entity, although in earlier chapters a degree of focus was given to individual and important players such as Japan, Singapore and Hong Kong. It is now necessary for a more detailed approach to be made toward the role of the People's Republic of China (PRC), whose spatial geography is of physical order that permits that country to be a major influence on the aviation industry both regionally, nationally and internationally at all times.

As an emergent economy of very significant size, the PRC plays multiple roles, not least in the economic futures of the significant number of countries that share a contiguous border with it. We also have to bear in mind that the existing arrangements with the United States have a competitive counterweight in the determination of the EU to engage in its own form of mutual relationships with China. From the perspective of Beijing, the visits of President Hu Jintao to many and various countries in recent times is also indicative of the growing interest of China in building international relationships as an emergent superpower. While it is important to be mindful of this larger geopolitical picture, it is now time to attempt some analysis of China's role as a key regional player, especially within the developing network of airports, of which it, of course, has the largest share from a regional perspective.

The Significance of China's Emergent Role as a Major Aviation Transport Market

It is important to remember that China has been in the process of evolution, from a centrally planned command economy to an increasingly liberalized and open market environment, for the last 25 years. From a political perspective, as perceived by Deng Xiaoping, China is now being led by the fourth generation of leadership, which began with Mao Zedong. It is also firmly committed in the new generation to Deng's famous strategy of 'building socialism with Chinese characteristics'. As a consequence, all activities involving regulatory and enterprise reform have a strongly pragmatic and problem-solving intention. China has also, as past events have shown, flavoured this with a large component of gradualism and a significant degree of privatization in specific industrial cases.

Prior to 1980 (Zhang, 1997), China's aviation industry, which had been founded in the early 1950s, was a quasi-military organization the Civil Aviation Administration of China, (CAAC), which operated as a department of the Chinese People's Liberation Army Air Force. As a result, it was driven by an administration system that operated at four levels. With Head Office located in Beijing, civil aviation was driven by six regional bureaux, 23 provincial bureaux and some 78 civil air stations. The entire system was micro-managed directly from Beijing, as a total monopoly. In addition, there were regulations covering market entry, route entry, frequency, pricing and passenger eligibility for air travel.

Needless to say, between 1953 and 1978, it suffered annual losses, even where government subsidies were factored into the balance sheet. As an example, the 6 years between 1968 and 1974 alone (Shen, 1992) saw 360 million yuan written off. In reality, there had been little or no development of a domestic market, since air transport was simply perceived as primarily a tool to be used for administrative and government convenience.

The Two Stages of the Civil Aviation Reform Process

The initial stage of reform saw CAAC separated from military control and management. Government at this stage concentrated on the transfer of responsibility for efficiencies in economic performance to the airlines themselves. At the same time, CAAC did retain a significant control over operation as well as airports and other service functions.

Beginning in 1987, a second stage of reform policy was aimed at separating the regulatory role from the operational function, which led to the airlines becoming profit-seeking units. The airports also benefited from the government's intention to separate airport operations from airline operations, while at the same time decentralizing the airport system. A pilot exercise at Xiamen airport, which served a major free enterprise zone, saw control transferred to the municipal government over all the fixed and working capital of the entity as well as its personnel. The system was further extended to the existing site over a number of years. Current policy requires that new airports be managed by local government from the day that they open.

Further liberalization included the easing of limitations on new carriers, which has led to a significant expansion of local airlines. By 2001, the total number of routes had reached 1143, of which 1009 were domestic schedules. These linked 130 cities within China, while Air China serviced 134 routes connecting main hubs, with 62 cities in 33 countries. The early years of the current decade were not a good time for the Chinese airlines. The impact of the financial crisis in the region, together with the SARS epidemic, saw the existing and relative inadequacy of domestic demand for air services become exacerbated. By 2003, however, strong growth in passenger demand was reported as a consequence of China's entry into the WTO and the remarkable growth of internal tourism.

To this may be added the growing size of inbound tourism, which saw revenues top $US25 billion. As an indication of the sheer size of the growth trend, the civil aviation sector made a profit equal to US$1.04 billion, which equals the aggregate sum of profits between 1997 and 2004. The period to 2004 was also remarkable for the further move by government to implement major changes in both the airline and airport sectors.

In July 2000, the reorganization of ten major airlines under the control of CAAC into three distinct air transport groups was authorized. From this decision there have emerged the current major airlines: China Southern, China Eastern and Air China. While the latter remains the flagship for international services, both Southern and Eastern are pioneering their own off-shore routes. Inevitably, all of the good news has a downside. The growing momentum driving demand raises, yet again, the important question of China's ability to resource and manage such growth. One major response can be found in the regulative changes on investment issues on 21 June 2002.

The Move to Increase FDI in the Civil Aviation Industry of China

Foreign investment in China's civil aviation sector is not new, as the frequently remembered 1995 investment of $25 million (14.8 per cent) by George Soros in Hainan Airlines will attest. The intention is to not only seek direct FDI. In addition, an externality is sought in the form of western-style advanced management techniques and models. Investors are sought under the regulations for airside as well as landside projects. Following to some extent the American method, a 'relative majority share' must be held by a Chinese party, which means that the indigenous holding as a percentage must sum to a figure larger than that of all FDI holders in the project combined. These requirements constitute the first category of the regulations.

The second category relates to investment in domestic airlines and caps one FDI initiative at 25 per cent, while the aggregate of all FDI in a given venture may not exceed 49 per cent (formerly 35 per cent under the 1994 regulations). The distinction is also made between governmental, industrial and tourism services, which must have a majority of Chinese investors. On the other hand, a majority of the equity can be held by foreign interests in agriculture, forestry and fishing ventures.

This latter condition is very interesting, since it seems to address a major Chinese problem, the markedly different levels of development that have occurred in the key urban regions, such as the coastal and southern regions, and the more backward and remote provinces of the country, in the west and northeast. It remains to note that new firms created as a result of the increase in FDI must have legal status as a Chinese business. On the other hand, there is no restriction on the grounds of citizenship, with regard to any post in a new venture, which includes appointments from the post of chair downward.

The Transfer of Airports to Local Administrations

Under the Civil Aviation Reorganization Plan formerly adopted by the State Council on January 22 2002, control over 129 civil airports directly managed by CAAC was transferred to local governments throughout China. The exceptions were Beijing International, Baiyun International at Guangzhou and the civil airports of Tibet. The decision raises significant problems for the new operators, since they are required to both finance airport development and find capital to cover the costs of operations. The situation is particularly fraught for the western regions, which are already targeted for developmental aid by government. How far Beijing will go with support for these regions remains to be seen.

On the other hand, the last 3 years has seen airports seeking the kinds of financial instruments that they need, through indigenous financial markets. It is clear that the growth of China's nascent bond, stock exchange and industrial and business financial markets will become an increasingly important factor in the provision of development finance in the future. In addition, several of the key hubs are commencing the processes for a change of status to that of limited liability companies. The lead has been taken with Shanghai Pudong, Kunming and Beijing Capital out in front.

As airports mature in their new roles, there remains also the possibility that major players such as BAA, TBI, Fraport and other managerial specialists will be drawn into the market by what appear to be significant opportunities for their own growth and development. Such possibilities, however, must inevitability be subject to the degree and extent to which central government will allow market liberalization to proceed.

Another important variable in the policy mix is the fact that as a member of the WTO, China must now satisfy the terms and conditions that apply under international regulations. As a working example, the current policy adopted by Beijing, which aims to create bilateral trade agreements, must inevitably meet the requirement of the standard most favoured nation (MFN) clauses, as laid down initially by the GATT and now by the WTO.

In general terms and from an aviation perspective, there can be no doubt that China is setting the templates for emulation by other countries in the region as potentially the largest aviation market in Asia-Pacific. This is a non-trivial assumption, which emerges from the very nature of is spatial geography and national endowments.

The focus of discussion has in a real sense assumed, in keeping with the vast academic and professional literature on aviation, that the basic strategic necessity is to respond to the inevitable expansion and growth of the aviation industry on a global basis. While there is an inevitable logic in this assumption, which gains support from the practical evidence of day-to-day operational activity, there remain increasingly manifest problems and uncertainties, which must come into any discussion of this kind. Of these, the paradoxical gap between rising demand and the seeming inability of legacy carriers to make a profit looms large. In turn, the material success of the low-cost revolution is beginning to be compared with the relative degree of 'service saturation' now found in some key markets.

From a macroeconomic perspective, the question of oil prices looms large, exacerbated by both environmental and geopolitical concerns. From a geopolitical perspective, as further discussion in the next chapter will indicate, a degree of tension is emerging as the United States continues to post extremely large deficits on the current account, the dollar weakens and pressure mounts on China to revalue its currency. In addition, inflation is picking up and credit card debt is increasing as consumers maintain spending levels above their incomes.

Despite these growing international uncertainties, the global economy remains buoyant. The problem is a question of time. How long will it take for the patterns of internal consumption to adjust to declining real productivity and, with it, a rising current account imbalance? Again, what will happen as the costs of oil continue rise and airlines attempt to offset this trend by setting premiums on tickets, especially on long-haul traffic? This latter question is especially sensitive, given the fact that it appears that competitive market pricing to limit the challenge of LCCs is already an active component of airline strategies.

It is possible to extend these speculative questions to include the airports, especially on the question of the control and management of slots. Expectations that a strong rise in the demand for slots as a function of growing demand will allow the regional hubs to raise fees especially to new entrants looms quite large. This already raises the question of the primacy over operational schedules exerted by national flag carriers. Can this be maintained if government also insists that slot usage for national airlines be subsidized?

The level of uncertainty in these matters is raised further if, as a consequence of a continuing rise in the price of aviation fuel, major foreign carriers decide to limit services to only those routes that can sustain some set target of capacity in the medium to long term. If we add to the mix the effects of terrorism, such as the recent Bali incident, or natural disasters, such as Hurricane Katrina, on the tourism markets, the risk factor very clearly will need constant evaluation.

The very large questions that arise in terms of the impact of these events on the aviation industry in general, and the airport industry in particular, will form the basis of the next chapter. There can be no doubt that the macroeconomic issues raised above will inevitably exert their influence in the future. At the same time, any judgements as to the real effects of changing circumstances will need to be somewhat speculative, as the issues raised in the next chapter will indicate.

Chapter 8

The Future for East Asian Airports:
A Speculative View of Emerging Problems

Introduction

The main purpose of this chapter will be to attempt some evaluation of a range of issues that are in the process of emergence, which makes them naturally subject to a degree of speculative consideration. The choice has been shaped by the large geopolitical events that will inevitably impact upon the aviation industry and, by definition, the world of the international airport. In turn, discussion will also focus on a number of issues that are beginning to be added to research agendas, by both academics and managerial think tanks.

The reason for a speculative approach to some of the issues that will be discussed arises from the fact that some of the evidence that is arousing intellectual interests remains subject to alternative forms of interpretation. From a research perspective, the basic cause may be found in the fact, that in some cases the assumptions reported are subject to predictive, rather than experiential data.

It is also important to note here that much of the economic information that will be evaluated below is shaped by the need to interpret market signals for evidence of their effects, as forces that will affect the international aviation market during the balance of this decade. Needless to say, given the current uncertainties facing the global economy in general, a significant degree of extrapolation is necessarily involved.

The primary intention of this chapter is to review a sample range of such issues, moving from the impact of changes in the direction of global economic growth, through to significant technological, managerial and operational themes that are currently matters of no little controversy in the aviation industry. The direction of discussion will move from the international to the regional. This is to allow for an attempted evaluation of the possible influences of issues and events on the East Asian region and its international airports in particular.

The chapter will begin with a consideration of the conflicting evidence that is emerging over the current prospects for global economic growth for 2006 and beyond. It will consider in some detail the central role of the United States as a primary influence on economic progress, within the context of rising concerns over inflation, the price of oil and the possibility of a rising political pressure on China to abandon the comparative advantage it enjoys in its export sector, as a result of a clearly undervalued national currency.

The focus will then shift to a consideration of the possibilities that might emerge for the East Asian community within the larger APEC context, if momentum picks up towards a formal FTA in the Asia-Pacific region. The possibilities of either a major conurbation emerging along the lines of the EU or NAFTA, as against a set of sub-regional arrangements, will also be canvassed, bearing in mind the centrifugal pressures that make East Asia the natural centre of the Asia-Pacific region.

There will be a very significant change in the direction of discussion at this point, with attention turned toward the future roles of the airlines. The longevity of the legacy carriers remains the focus of increasing speculation, for reasons already considered in earlier chapters, and a further evaluation will be attempted. In turn, the large questions facing the future of the low-cost sector will be considered, with attention focused on the large question of the emergent need for consolidation and some inevitable form of market clearing.

The advent of a new order of aircraft designed for ultra long-haul and non-stop services at a premium will then become the focus of attention. The contentious questions being raised with regard to the degree and extent to which political subsidies distort the competitive market for new designs will be reviewed. In addition, consideration will be given to the attendant emergence of two new consumers of long-haul non-stop services. First, the price-indifferent consumer of the new premium and customized services will be considered, followed by the new order of expatriate business travellers, whose terms of external appointment are getting shorter and subject to increasingly common travel between the home office and the external location.

The direction of discussion will then attempt to consider the implications of all the previous themes, for the international airport and from a developmental perspective. The need to balance existing hub-and-spoke services with the exigencies imposed by new mega-carrier point-to-point arrangements offers an important example of this theme. The chapter will then close with a review of the issues raised throughout the discussion and the possible directions that might emerge from the necessary strategic decisions airports will need to make to accommodate such changes.

The World Economy: Rapid Growth versus Increasing Instability in 2006

Analysts and other observers of the state of the global economy appear to have reached a degree of consensus on the view that while the international economy is still in a rapid growth phase, there is a significant degree of instability emerging (*Economist*, 22 September 2005), which raises the risk factor in making forward predictions. According to the IMF, the growth rate for 2005–06 is 4.3 per cent, well above trend. Unfortunately, there are now signs that factors such as the strong surge in oil prices, coupled with the effects of Hurricane Katrina and the fact that the US external deficit stands at 6 per cent, are beginning to indicate a slowing in the global economy (Mussa, 2005). At the same time, it appears that the world equity markets do not yet reflect that possibility. Put simply, the markets would appear to have a significant form of lagged response at the current time.

On the inflation front, the sharp rise in the price of energy is finding its way into consumer prices, although the degree and extent to which this is becoming a permanent effect is hard to judge. The vital health of the oil industry is clearly high on everyone's agenda, particularly in the light of the national disasters that have struck the United States. While there is a salutary need to remember that oil prices were rising before Katrina, any attempts to predict a return to normality in production for the regions worse affected will need an instant correction. This is because her sister, Rita, due at the time of writing, will move inevitably through the key Mexican and Texas areas.

What does appear to be evident, with regard to the medium to long term, is the fact that there has been a major lift in world demand, which has to be set (Mussa, 2005, p. 3) against current limitations on the ability to expand supply. In effect demand growth exceeds supply growth the need to dampen demand and how to do so, thus becomes an important strategic question.

These problems have emerged at a time when the United States as the primary western economy faces internal problems. The external deficit has been widening for a decade, and there appears to be a real need to moderate such factors as internal domestic consumption through household spending and the current need (Baily, 2005) for a softening of demand in the domestic housing market.

Such measures are primarily being influenced by the fact (*Economist*, 22 September, 2005) that the US Federal Reserve has tightened its funds rate, despite expectations that it would not do so for fear of a slow-down in economic growth. The reason can be found in the sharp upward movement in the rate of expected inflation. According to the very influential University of Michigan Consumer Index, their recent survey indicated that consumer expectations anticipate a sharp rise in inflation caused by an expected rise in the price level of 4.6 per cent in 2006. This, in turn, reflects a predicted rise in the core inflation rate from 2.8 per cent to 3.1 per cent over the next 5 to 10 years.

It is worth noting at this point that firms are already passing on some of their higher fuel costs. Both airlines and delivery firms have raised their ticket prices. In the legacy carriers' case, the addition of a surcharge to the ticket adds a further counter-competitive burden to an already severely price-disabled industry. Further complications arise because businesses are now facing a period of rising wage pressures. Unit labour costs rose by 4.2 per cent in the year to June 2005 at a time when productivity was falling. It is a matter of no small importance that the same trends also appear to be emerging within the EU and at the same time.

Further discussion of the technical aspects of the current situation really lie outside the scope of this chapter. At the same time, the implications of the current uncertainties do have serious implications for the growth and integration of East Asian economies, with an inevitable flow-on effect into the aviation industry. Given the major developmental role expected of the international airports in the region, the implications are very serious, as discussion below will indicate.

The Implications of a Downturn for the Key Asian Economies

The focus of discussion thus far has centred upon the effects of current macroeconomic uncertainties, upon the United States and Europe. It is now time to consider the impact that the downturn will have on the East Asian region, bearing in mind from previous commentary that important issues, such as the availability and flow of FDI, might well be seriously affected.

There have been growing signs in the case of Japan that over a decade of economic stagnation after the bubble economy of the 1980s is coming to an end. This view seems to be endorsed by the recent success of Prime Minister Koizumi, whose mandate to privatize Japan Post received a strong endorsement in the recent election. On the other hand, as a country conspicuously lacking in natural resources, the sharp rise in oil and its negative effect on energy prices will have a serious effect on both business and consumer confidence.

By contrast, China faces a number of serious problems, but of a different order. It has been for some time subject to the effects of a growing economic surplus due to the growth of exports. In addition, it has been the primary global focus of net inflows of FDI. This has led to recent moves by central government to slow internal investment as part of an attempt to control rising domestic demand.

The problem with this strategy (Mussa, 2005, p.13) is that slowing growth in domestic demand coupled with moves toward import substitution in domestic markets have been upset, on the export side of the equation, by a strong improvement in the trade balance. China's successful emergence as a major exporting economy has been based on the fact that the national currency has had a fixed exchange rate, because the capital account has been closed. In other words, unlike most currencies, the yuan is not allowed to float internationally, which means that the export industries enjoy advantages both in terms of their internal costs and against their final export market prices.

How long China will be able to maximize the benefits from its undervalued currency is now a matter of serious contention, both in the United States and in Europe. This is reflected in the events that have surrounded the removal of the international regulations that, until 2005, set production quotas for textile and apparel exports under the GATT.

The formal ending of the International Multifibre Agreement on 1 January 2005 formally removed quota controls over the production of textile and apparel products by exporting countries such as China and India. These had been originally intended to prevent monopolistic competition between, amongst others, transnational corporations. The conclusion of this arrangement has given China the role of market leader, able to exploit its comparative advantage both in terms of capacity and costs to fill the role of global leader in the industry.

This has led to a major reaction in both the EU and the United States. Under the terms of China's membership of the WTO, the United States and other member states with an industrial interest reserved the right to unilaterally impose quotas where their domestic industries experience disruption as a consequence of a surge in foreign imports. This led to a ban imposed by Belgium and France on Chinese clothing arriving at their ports of entry, which has been partially removed pending further action by the European Commission.

In turn, the United States government, in May 2005, re-imposed under the WTO regulations a series of quotas in Chinese cotton garments, on the grounds that since such exports had reached 80 per cent by volume, there was a need to protect the interests of America's indigenous apparel industry. The contents of the embargo have since been extended to cover other forms of apparel. Ironically, the views of the textile manufacturers were not shared by the retail trades, who predicted that US consumers would face higher prices.

At the time of writing, these issues are subject to ongoing discussions between trade officials from both countries. According to the US Committee for the Implementation of Textile Agreements, nothing short of a comprehensive agreement will see the embargo raised. There can be no doubt that high on the agenda will be the question of the undervalued status of the yuan, if only because it has been a target of US complaints for some considerable time.

The current uncertainties already discussed above will add to the pressures on the Chinese to let the yuan adjust upward to what the market will decide is its true value. If discussions over the balance of 2005 and beyond do not progress this item as number one on the agenda, it is possible that a serious clash will occur between the United States and China. This may take the form (Cline, 2005) of the establishment of countervailing duties against China, on the grounds that failure to appreciate the value of the yuan will be an act in contravention to both the GATT and the IMF requirements. In the first case, the failure to act runs counter to commitments to free trade. In the second, it falls foul of rules against the manipulation of exchange rates.

The economic and geopolitical circumstances discussed above have been intentionally detailed and extended, in proportion with the possible serious consequence for the East Asian economies, and the larger Asia-Pacific region in particular, which in total encompasses both China and the United States. It is now time to address the issues within that geographical context.

The Further Potential for an Asia-Pacific Economic Crisis

In a very real sense, the increase in geopolitical tension may be seen as two-dimensional. On the eastern side of the Asia-Pacific region, the United States has a global current account deficit of almost US$800 billion, which is 7 per cent of GDP. It is rising at the rate of US$100 billion per annum and the country must borrow externally at the rate of US$6 billion per day. At the end of 2004, the net foreign debt of the United States stood at US$2.5 trillion and it is increasing at the rate of 20 per cent per annum. While the aggregate level of debt in unsustainable, the growth trajectory is even worse.

Using an aviation analogy, its like an airliner in the climb out from a major airport: unfortunately the captain has no idea as to the designated cruise height, which means that a very urgent search is now underway to find one. The stark reality of the situation requires that annual debt must be cut by at least 50 per cent so that the ratio of US foreign debt to GDP can be stabilized. Failure to do so will place the sustainability of the entire economy in real doubt, unthinkable as this may be as an outcome.

Meanwhile, on the western side of the Pacific, which of course includes East Asia, the opposite situation prevails. In a very real sense, China is something akin to the biggest kid on the block, which with the Asian countries included, has accumulated massive foreign exchange reserves, accounting for 90 per cent of the global increase between 2003 and 2005. It should be no surprise, therefore, that much of the borrowing by the United States comes from the countries of the region, notably led by China, Japan, Taiwan and Korea. In effect, it has been suggested (Bergsten, 2005b) that the US deficit and the Asian surpluses are really 'two sides of the same coin'. Neither can make the necessary corrections in order to return to some degree of stability unless a debt reduction on the American side is accompanied by a surplus reduction on the Asian side.

In the meantime, the situation remains tense and the signals emanating from the United States indicate a movement towards further stiffening of punitive measures. For example, should China delay the appreciation of the yuan, it is likely (Bergsten, 2005a) that the US House of Representatives will pass the Schumer Amendment, which will place an across the board surcharge of 27.5 per cent on all Chinese imports into the United States,

There are also indications that the United States, which pioneered bilateral aviation agreements, is now intent on a parallel development of substantive FTAs. President Bush is expected late in 2005 to initiate discussions for an FTA with Korea. Strategically, this will place Japan at risk from trade diversion, which could lead to a Japan–US FTA. Given the fact the America is simultaneously launching talks with several countries in Southeast Asia, the risk that China will see this as a form of containment strategy looms large.

The effects of these developments on the economic stability of East Asia have clear ramifications for the aviation industry in particular and inevitably to the evolutionary future of the key international airports. The least damaging strategy would appear to take the form of an all-embracing FTA that would expand the membership to include the South Pacific island states and the CER countries of Australasia. The matter will remain urgent and with the United States also engaged in discussions for a FTA of the Americas, the possibility of a political line being down the middle of the Pacific could have profound consequences for any multinational open skies agreements in the future.

The direction of discussion now moves away from the consideration of the macroeconomic and geopolitical issues that will inevitably do much to shape the future of the East and Southeast Asia region. What follows is a review of some further emergent issues, this time within the context of the aviation industry. During the earlier chapters, significant attention was paid to the future roles and scope of the international airport as the key sector under review. Attention will now be turned to some emerging issues on the airside. These will be examined against the background of earlier discussion. The ensuing discussion will also incorporate some of the pioneering research work that is emerging, as interesting problematics come out of the current confusion and flux in international aviation.

Do Mainstream (Legacy) Carriers Really have a Sustainable Future?

Prior to opening the discussion, there is a need to give a degree of precision to the term 'legacy carrier'. It literally refers to those American airlines, United, American and Delta, whose antecedents were created by formal government regulation in the late 1930s. In more recent times, the term has become synonymous to some extent with the word mainstream, essentially those airlines that constitute the major operators of a full range of in-flight service in most countries, with a status ranging from national flag carrier to major international player.

There is a significant degree of ambiguity apparent in the fact that some national airlines are, in ownership terms, to be found in various degrees of privatization. What becomes important at this point is the degree of limitation involved, ranging as it does from 100 per cent public ownership to the same amount of private control. The problem with a continuing national identity, it has been suggested (Lelieur, 2003), is the fact that today's mainstream carriers are an essential part and embedded in the larger operational context of global business, be it tourism or other forms of multinational expansion.

The American experience remains the most obvious example of the problems, both internal and external, that face the airlines whose mandate is to operate both a domestic and international service. It appears that despite the persistent decline of the legacy group en masse, there remains significant and positive scope in international services. United, for example has established a strong position for potential inter-hub services between the United States and China; in addition, it has established a presence with its own booking office in Guangzhou.

On the adverse side, it is clear (Williams, 2005) that federal regulations with regard to both foreign mergers and FDI impose a serious limitation on the major carriers, which would not be tolerated in other business sectors. In the first case, federal anti-trust law allows challenges to be mounted by the Department of Justice where the merger might offer the parties a potentially monopolistic opportunity. As a working example, the takeover of TWA by American in 1991 was challenged on anti-trust grounds. The action failed and ironically American later closed TWA.

In the matter of FDI, the law imposes a minority shareholding restriction on the proportion of equity that might be obtained by overseas interests. This requirement has an obvious and deleterious effect on the ability of a given airline to raise needed capital. Meanwhile, the current and prolonged crisis has had serious effects on the market capitalization values of all the legacy carriers. These are now rated BBB by the money market, which reduces the value of shares in all of them to the rank of junk bonds.

The problems of access to fresh capital resources and the constraints on mergers imposed by anti-trust law have literally left the major carriers in the United States, as the *Economist* recently commented, 'flying on empty'. This was somewhat dramatically demonstrated in May 2005 (Williams and Williams, 2005), when the Federal Bankruptcy Court, on application by United Airlines, allowed management to default on four pension plans, which covered some 120,000 retired workers and 62,000 active employees. What is remarkable about this event is the fact that United sold 55 per cent of the company to several groups of workers, both unionized and non-union staff, under an employee stock ownership plan (ESOP) in 1994. It remains to comment on the final irony that even if the right to merge was allowed in law, given the mountain of debt facing all of the major carriers, their managements would be exceedingly wary about either mergers or takeovers; they might inherit it in the process.

These sad events appear to be scheduled for further repetition in the matter of rising costs. Scarce oil refining capacity in the United States requires that aviation fuel production competes with other market products for available output. When this problem is coupled with the current spate of hurricanes striking key sites, it could leave jet fuel at a final price to airlines of US$87 a barrel (*Economist*, 22 September 2005).

As to the future of the full-cost carrier, there appears to be little to look forward to, at least in their current organizational formats. The most successful sub-sector of the industry, that of international routes, is beginning to attract strong feeling in Europe over the status of Chapter 11, which appears to be devoid of any disciplinary conditions attendant upon discharge from that status. While it is suggested that the future may lie in the further development of current strategic alliances into more formal and mega operators, the grand strategy to bring such development to fruition appears to be still some distance away.

There remains, of course, the final question: how then will the full cost sector of the airline industry respond to what is an undoubted and growing demand for air travel? The need for an answer is particularly germane in the case of the East Asian region, since the demand can be related to the emergence of a new and growing middle-class society in China and the other ASEAN states. How the sector will respond to what are really structural changes in their conventional market will shape the organizational future of what remains a key part of the airline industry.

If the future of the legacy carrier is in some doubt, the same questions might well be asked of the low-cost variety, as start-ups expand and develop on an increasingly international basis. At the centre of what is an emerging debate are questions relating to sustainability and the possibility that the market might well, with increasing maturity, start to see inefficient carriers leave and new start-up decline.

Will the LCCs Achieve Real Sustainability?

The LCC sector has found itself in the vanguard of the airline industry, in the sense that it has been able to tap into and maximize a previously unrecognized demand for air travel. It has done so by offering low-cost and no frills services, which have been taken up by segments of the population that had not previously been seen as clients by the industry. It was also able to maximize interest from business travellers, especially those whose primary demand was less for low-cost and more for point-to-point speed to their required destination.

With some of the icon airlines exhibiting a growing maturity, an important question arises. Given their strong emphasis on low costs, how vulnerable are they to changing circumstances? In other words, does their growth strategy include a long- to medium-term perspective? A recent and comprehensive research project (Gordon et al., 2005) revealed that there was a distinct lack of a strategic management strategic perspective to be found within the sector. In addition, there was little understanding of sustainability issues. Of equal significance was the lack of an understanding of key issues, such the possibility of potential market saturation, together with the consequential need to sustain a competitive position. Other matters, such as potential liabilities that might be incurred in relation to flight safety and the increasing attention being paid to environmental issues, were also subject to lack of awareness. In sum, the dominant stress was on market growth and expansion.

American studies have also lent endorsement to the view that at some time in the future there will be a significant shake out in the US LCC sector based on a potential glut of seats, which will simply leave LCCs without a cost advantage. This view is endorsed by Spanish research, which notes that in the European case, the market is now dominated by three major players, Ryanair, Easyjet and Air Berlin, fourth place taken by Hapag-Lloyd. Together they have just over 50 per cent of the European markets, which comprises some 48 separate airlines. A change in pricing over time is also noted, with the anticipation of that there will be some price rises as some airlines struggle to meet rising costs. In addition, there are clear signs that as the main players consolidate their market shares, less competitive carriers, as in the case of the Italian company Volare, simply have to leave the market.

While Southwest continues to evolve into an operational status somewhere between its origins and membership of the full-cost sector, the British carriers Ryanair and Easyjet have sought new revenue streams from the sale of advertising on the backs of seats as well as on the aircraft exterior. In addition, they offer car rentals, travel insurance and travel reservation services, with all of these activities accounting for some 10 per cent of their revenue streams. In addition, there seems to be talk in the industry about the possibilities of expanding revenue streams to include in-flight activities such as gambling and other forms of entertainment.

With the exception of air and cabin crews, reservation agents, some head office functions and some maintenance functions (Gillen and Lall, 2004, p. 47), all other activities are outsourced. The sector is notable for the relative absence of employee organizations for the range of occupations that are active both on the ground and in the air. This has allowed the LCCs to set remuneration rates in which the proportion of basic pay to incentives-based income is much smaller than mainstream airline rates. In the case of reservation agents for Easyjet's Luton hub, it tends to employ teenagers who are paid solely on a commission basis. In fact, productivity and commission constitutes the entire reward system.

The question inevitably arises as the airline moves into organizational maturity, will those workers who have been retained over considerable time begin to look for compensatory and service-based arrangements? In other words, will the LCCs, lose their relative advantage over the full-cost operators with regard to their labour costs? The alternative would be to retain the current system and in the absence of a real human resource management strategy, absorb the high costs of labour turnover as well as the opportunity costs that arise due to inevitable inefficiencies in the operational network.

It is clear that these questions are not lost on the market leaders, as the comments made by Ryanair's Chairman to a recent major industry conference makes evident. He noted that as new markets open up they will be developed only by the leading-edge airlines. He anticipated that, as the cost of oil rose, their market would begin to clear as weaker LCCs collapsed. Major concern was also expressed that new and intentionally constrictive regulations were beginning to appear, commencing with the now operational passenger rights legislation and restrictions on route development, the subsidization of carrier by regional airports and increasing concern over aircraft emissions. It is quite clear that the LCC sector in Europe is facing up to a significant series of market adjustments, as rising costs allied to legislation aimed at regulating the sector begin to bed in over time.

The Current Emergence of Discount Airlines in Asia

The arrival of the LCC has been a significant event in the aviation industry across Asia. Its arrival was signalled with the launch of Air Asia, a Malaysian airline that was launched by Tony Fernades in January 2002. The airline had had a bad start after its original foundation by the government-owned conglomerate, DRB-Hicom. It was heavily debt laden, when, in a symbolic gesture to show that Fernades owned Tune Air Squadron Berhad, he paid the government a price of one ringgit or 25cents.

The airline has now grown into a very successful and profitable business. Strategically based at Senai International Airport in Johore, it now flies throughout the Southeast Asian region as well as to Macau, Xiamen and from Singapore, just across the harbour. True to the Southwest tradition, it operates a no frills service based on open seating and Internet bookings.

Since Air Asia arrived, it has been joined by some 25 new players. The general consensus is that not all of them will survive. While the common strategy has been to build a clientele in the market as quickly as possible, intense competition coupled with high fuel prices has made things difficult, especially for the very new entrants. The major problem at the moment appears to be on the supply side, with the possibility that there is a current asymmetry between the current level of demand and the number of airlines seeking passengers.

There is also a waiting list of new applications to launch, which makes the demand for some degree of consolidation something of an imperative. Meanwhile, Singapore is acting as the major hub for the region. At the same time, there can be little doubt about the market's future when it is recalled that there is a burgeoning middle class to be found in all of the countries across the region, which now have access to low-cost and efficient cross-border transportation.

The expansion of LCCs has also reached India, where the airline market is going through a liberalization process. To date, Air Deccan has operated a domestic network, but the distinguishing factor of spontaneity of purchase is not present due to the fact that seats have to be booked often several months ahead. Air India Express, by contrast, is the LCC spin-off from the national carrier, Air India, with cross-border links into the Middle East, particularly the gulf states of the United Arab Emirates. The launch of Kingfisher Airlines, which is owned by India's leading beer and liquor tycoon, not only carries the name of his most famous brand, but also promises fashion models as hostesses and in-flight entertainment at every seat. These new entrants will inevitably face a challenge from the East Asian group of airlines, as bilateral agreements open the doors to cross-border services. In fact, Valujet has already commenced a Singapore to Calcutta service.

The extent and range of carriers that currently compete in the East Asian market are shown in Table 8.1.

The types of services offered by the airlines range from in-country turbo prop travel, to destinations within the Philippines by South East Asian, to modern jet services connecting cross-border city pairs within the capitals and other major destinations in the ASEAN countries. The market strategies adopted by the new services seem to be very much subject to governmental perceptions of their value.

For example, Japan remains firmly committed to city-pair services within its national boundaries, which places the LCCs in direct competition with the major airlines, whose 747 internal shuttle services command a mass market. By comparison, servicing cross-border networks within the Southeast Asian group of countries appears to be a common goal of 14 airlines currently operating in and out of the region.

The precise future of the low-cost sector would seem in the final analysis to depend upon a problem that was discussed in an earlier chapter. It refers to the fact that an ASEAN-wide open skies agreement is still in the discussion stage. The rapid growth of the LCC market in Europe was made possible by the full cabotage conditions that followed the introduction of the 'third package' of liberalization reforms in 1997. How soon the ASEAN states agree upon and confirm their own open skies arrangements would appear to have a very important bearing on the future of the LCC in East and Southeast Asia. In the meantime, passenger demand continues to rise and the demand of new operators to be allowed market entry continues to grow.

Table 8.1　　The operational distribution of LCCs in Asia

Country/region	Name
China	Spring Airlines
Japan	Air Do
	Skymark Airlines
	Skynet Asia Airlines
Korea	Hansung Airlines
	Jeju Air
Southeast Asia	Air Asia (Malaysia)
	Air Paradise(Indonesia)
	Asian Spirit (Philippines)
	Cebu Pacific Air (Philippines)
	Jetstar Asia (Quantas/Valuair Singapore)
	Indonesia Citilink
	Lion Air (Indonesia)
	Nok Air (Thailand)
	Orient Thai (Thailand)
	Phuket Airlines (Thailand)
	South East Asian Airlines (Philippines)
	Tiger Airways (Singapore)
	ValuAir (Singapore)

Source:　WikiTravel.com.

The Expansion and Growth of Long and Ultra Long-haul Aircraft Capacity

The development of air transport services over time, have always emerged from a symbiotic relationship between new aircraft technology and the airlines. With the advent of the wide-bodied Boeing 747, the potential non-stop point-to-point services became both technologically and financially viable. In addition, major advances in new aircraft technology, particularly since the mid 1980s, saw the introduction of the twin engined Boeing 767 and with it the emergence of the extended twin engined operations (ETOPS) system. This has allowed for a further expansion of both range and endurance, especially with the coming into service of larger twin engined aircraft such as the Boeing 777 series. These aircraft are now capable of very long range flights, free from the routing restrictions imposed by the ETOPS system on the earlier generation of 767s and their progeny.

　　Competition in the market has also been forthcoming from Boeing's great rival Airbus. On the other hand, to date, the company has put great reliance on its four engined A340-500/600 series, which have matched to some extent the range and distance of the 777 series. This raises again the interesting controversy between the two manufacturers as to the potential demand for new aircraft up to the year 2020. Both firms agree that the bulk of demand will be met by the supply of twin-engined types, ranging from single- to twin-isle configurations according to the service requirements of specific airlines. On the other hand, it is also interesting to note that, with the advent of the proposed Boeing 7E7 series, Airbus has countered with the A350, its own twin-engined variant.

The world now awaits the arrival of the Airbus A380, which in a very real sense, though the fact is denied by EADS, its construction company, is really a much bigger, version of the 747 genre, with, of course a longer range, in excess of 8,000 nautical miles. Such is the capacity of this three-decked carrier that it has been officially placed in a new category of aircraft types (VLA6). It is able in the first 800 series to carry 555 passengers in a normative three-class configuration, rising to 635 plus. Assuming a single-class configuration, this raises the maximum passenger load to between 800 and 900.

The development of the A380 marks a serious risk for EADS, the primary manufacturer of the A380 triple-deck jumbo carrier. It has already been remarked that Airbus quite clearly sees that the paramount demand from new airlines and established carriers seeking fleet replacements will largely be found in the various twin-engined categories. At the same time, Airbus would argue that there are serious problems, such as major congestion at key hubs in many countries, which is compounded by the major lags in airport development that is required to meet future demand. The arrival of fewer aircraft, as exemplified in the A380, coupled with an increase in carrying capacity might go a long way to solving congestion difficulties.

Such an assumption, however, raises a second order of questions. In order to maximize the promised efficiencies from purchase and operation, the new order of mega-jumbos will need to fly point-to-point. This requires that airports manage not only their network banks, but also the space-time and development costs of servicing these new aircraft. This has been a notable problem for US airports, since they remain very largely in public ownership, or subject to restrictions on new projects, especially where a dominant airline has a leasing clause that requires consent for such work.

Assuming that the airlines that are scheduled to acquire the 380 maximize its potential for ultra long-haul service, the question immediately arises, who apart from economy class passengers will fly in the 380? The question is important because a number of airlines have already introduced long haul non-stop schedules.

The most obvious example is Emirates, which offers a 'total experience' to its first and business class clientele. It must be noted, however, that the overall game plan is to build a global service operating out of the super-hub at Dubai.

By contrast, several other carriers have been more precise in their definition of what seems to be a significant new market. In doing so, they have targeted as passengers business and other travellers, who are prepared to pay a significant premium for high-quality service in an aircraft suitably equipped to come up to their expectations. These include the need to travel very considerable distances quite frequently. Unfortunately, the conventional progression in doing so may often require a pattern of stops on the route in order for any given flight to fit into the bank requirements of a given international carrier's hub-and-spoke schedule.

Alternatively, it may mean interlining, with a change of aircraft. Further exigencies of seasonal and other forms of travel all combine to make a given long-distance journey both stressful and in some cases hazardous, especially if delays cause a knock-on effect and disrupt a passenger's planned itinerary.

Both Singapore International and Thai International offer regional examples of a new form of scheduling that attempts to integrate long-haul point-to-point travel with a range of services offered to a single premium class of travellers. The case example in Figure 8.1 demonstrates the Singapore International approach. It should be noted that in order to obtain permission to offer this service, the airline went through a stringent set of evaluations. These included carefully controlled testing in flight of such key matters as aircrew fatigue and related studies of rest periods and other support systems. As the popular advertising indicates, the notion of a personal environment for the passengers while on board is also carefully factored into the operation.

- Destinations: New York and Los Angeles (return service/point-to-point)
- Aircraft type: Airbus A340-500
- Passengers: 301 in two classes, Premium Business and Premium Economy
- Aircrew: Two Captains, two First Officers
- Cabin staff: 10-15 members
- Flying time: New York – 18 hours, Los Angeles – 14 hours
- Working cycle in flight: 19 hours (New York service)
- Rest periods: Two sleep periods are scheduled for all crew members
- Standard duty cycle: Two return flights a month

Figure 8.1   Singapore International Airlines: Premium services
Source: Airline literature reviews and personal interviews.

In the matter of passenger demand, this is reported to be exceptionally heavy, given the fact that the level of services permits travellers to continue with their conventional professional activities, while in flight.

International Business Travel as an Emerging Market

It is clear that the major carriers, who are targeting this particular market, appear to have tapped into an important market. As the following commentary will attempt to reveal, it might be much larger than is currently realized, and for reasons that will now be discussed in more detail.

The expansion of operational activities from the home country of a given corporation to a wide variety of international locations has been the core strategy behind the growth of MNEs on a worldwide basis and over the last 40 or so years. During that same period, it has been common practice for MNEs, including the major international airlines, to maintain the senior managerial functions in their overseas location by the traditional appointment of expatriate managers. Such positions tended to be of a medium- to long-term duration, with various contractual supports intended to alleviate both the social and economic costs of extended time away from the home country.

In recent times, the sheer cost of maintaining the system has increased. It has been further complicated by such factors as the rise of dual careers causing sometimes diverse expectations between partners and the emergence in some countries of indigenous, or even third-country candidates whose appointments might on the one hand alleviate the costs of expatriation, but whose formal preparation for senior posts requires a significant period of both in-house and external training. As a consequence, there has been a marked increase in what has been termed non-standard assignments (Welch and Worms, 2005). These include the types of contract arrangements listed in Figure 8.2, which indicate a range of different types of opportunities all with a significant international travel component.

1. Commuter: The manager engages in weekly or bi-weekly travel to a place of work in another country.
2. Rotational: The manager commutes from the home office to an overseas location for a set period, followed by a return and extended period of leave, before the cycle commences again.
3. Contractual: The manager with specific professional competences is assigned for short-term periods to a wide range of locations within the MNE network.

Figure 8.2   Types of non-standard international business assignments

These appointments are filled by designated managers who also fit into the category of International Business Traveller (IBT). In this category of business activity, their processes of work and travelling are essentially interlinked, or as one manager put it (Welch and Worms, 2005, p. 25), 'the travel is the work and the work is the travel'. The international literature on human resource management has tended to concentrate its attention on the long-term expatriate. In the process, that group of IBTs popularly known as 'globe trotters' or more accurately, frequent fliers, have tended to be somewhat ignored, as research has primarily concentrated upon the issues and problems arising from long-term expatriate appointments.

For the reasons already adduced above, the research and development focus within the MNEs' Human Resource Management (HRM) departments is now changing to accommodate the need to identify and support the ability of designated managers to commute between a number of sites, either regionally or on a truly international basis. Needless to say, such a shift requires the itemization of a new range of variables, where a manager's schedule might require visits to three or four countries within the span of not months, but rather a few days.

The logical vehicle that allows IBTs to maximize the values of time, distance and urgency is, of course, the non-stop flight in premium conditions that allow the traveller a number of advantages. These include the ability to continue carrying out various tasks, and increasingly to do so while remaining in real-time contact with office and home. Apart from an optimal physical environment, the IBT then needs to be able to choose between alternative activities as a means often of reducing stress and other problems associated with very frequent business travel. All of these elements will inevitably be factored into airline planning in the sure knowledge that competitors have also got their plans to obtain a share of what is a recognizably growing market.

Where Do the Airports Fit into the IBT Equation?

From an airport perspective, there is also good reason to establish services and facilities that are predicated on the first principle that, for professional IBTs, the journey really begins at the point of leaving home. Those airlines that now offer premium class flights only on very long haul routes do so with the purpose of making origin to destination conditions as seamless as possible. This requires not only the provision of services and facilities at the point of departure, but further links at the destination to allow for the further and smooth progression of the journey.

The need therefore to create forms of alliances between the twin hubs covered by the premium services clearly needs to be factored into the growth strategies of the major hubs. The ability to do so, however, would appear to be dependent upon the stage of development that aviation in general and individual airports in particular has reached. In Europe, for example, cross-border integration is well advanced. This raises the inevitable question, how ready are the major hubs of East and Southeast Asia to do the same?

Any formal conclusions would be premature in the extreme, given the various stages that questions of regional integration have reached amongst, for example, the national contributors to the development of ASEAN and APEC open skies agreements. At the same time, there are some indications that moves are underway, with the most notable example found in the East Asia Airport Alliance (EAAA), which began life in 2002 (see Figure 8.3).

What is really significant about the EAAA is not so much the size of the group, which incorporates some eight members, but the fact that they represent those airports that are clearly going to dominate the region, spatially, geopolitically and economically. They represent the dominant national economies of the region: China, Japan and Korea. Their current agenda encompasses a number of objectives that would enable the members to commence the development of an integrative relationship across the region.

The first intention is to develop an integrated website that will allow cross-border passengers, to obtain to obtain multi-language information at any airport that is a member of EAAA. This is to be further supported by the standardization of airport signage that will allow common pictograms and images to be used at all locations. Attention is also being directed toward the establishment of standards of service that are both of high quality and user friendly, with such images as 'Feel the Asian Way'. Finally, the EAAA has established an information sharing programme that will explore cost efficiency and profit maximizing revenue management.

1.	China	Beijing Capital Airports Holding Company
		Shanghai Airport Authority
		Hong Kong Airport Authority
		CAM-Macau International Airport Co. Ltd.
2.	Japan	Narita International Airport Corporation
		Japan Airport Terminal Co. Ltd.
		Kansai International Airport Company Ltd.
3.	Korea	Incheon International Airport Corporation
		Korea Airports Corporation

Figure 8.3   The membership of the EAAA
Source: Beijing International Capital Airport Newsletter, 6 June 2005.

If there is an indication of the future of the major airport hubs of East and Southeast Asia to be garnered from these developments, it is that the current distribution of geopolitical and economic power in the region is beginning to be imprinted on the airport sector of the aviation industry. Other major locations in Malaysia, Thailand, Singapore and the Philippines may well join the EAAA over time, but in doing so they will simply be adding to a development blueprint that is already in the process of being drawn up. At the same time, it is important to observe that a large number of conflicting issues and paradigms surround aviation policy making in the region, and they require to be brought into some degree of balance before real development can proceed.

The just completed commentary marks the completion of the final set of themes that have been the focus of discussion in the present book. In the best tradition of authorship, I began the research for this work, with a set of assumptions that I hoped, as a professional economist with a long-term interest in aviation, would simply be confirmed by the evidence. In the event, the wide range of complexities, both economic and geopolitical, have simply demonstrated the fact that a wider and deeper research agenda is needed to really capture the full dynamic complexity of the issues than can be encompassed in a single volume.

A final and brief set of comments will inform Chapter 9. They will deal with some of the portents of change that signal the increasing integration of the aviation industry into the larger contexts of international business development. In doing this, stress will be given to the fact that aviation is becoming more than a service industry and in the process merging into the larger and global contexts of trade and business activities.

Chapter 9

Reflections on the Future of International Airports in East Asia

Introduction

It is now time to briefly reflect on some of the many and various themes and topics that have interwoven into the discussions covered in the previous chapters. In doing so, the focus will be placed on some of the various aspects of the book, in which economic and geopolitical issues have predominated. The original intention when it was first conceived was somewhat tentative and deliberately described as a reconnaissance exercise. On reflection, the choice of approach was apposite, since the range and complexity of the issues facing the airport industry in the East Asia region remain somewhat daunting when the search for directions and outcomes is attempted.

What follows are some final comments and reflections on the widening range of changes that are now having considerable impact on the aviation industry in general and the airport sector, and in particular on their expanding landside activities. The speculative tone of the last will be maintained as uncertainties, such as terrorism, fuel prices, and pandemic diseases, come to the fore. What the issues do have in common, however, is the fact that they will continue to shape the future of the aviation industry as a matter of some urgency and over time.

Some Impacts of the Globalization of International Trade and Business on the Aviation Industry

The globalization of trade and business is having a significant impact on the aviation industry. The reason may be found in the fact that the changes that are consistently being imposed by the globalization of trade and business not only require the players in the aviation industry to be reactive to changing clients needs; they are themselves in their institutional and function activities being re-shaped and formed by the very same kinds of drivers that are influencing their clients. This is because over time, the aviation industry is becoming more and more integrated into the strategic and developmental activities of international trade and business. In fact, major airlines and international airports are by their operational identities now very much embedded as a matter of course in the planning of firms both within national economies and across external borders.

A very useful example of the proactive influences of change can be found in the aircraft manufacturing sector. Quite apart from the fact that there is considerable and increasingly political contention between the dominant duo of Boeing and Airbus over the question of government subsidization of projects, which continues to smoulder with the occasional eruption on either side, both companies now follow a pattern of production planning, which locates specific and technical aspects of their manufacturing cycle across a wide range of countries. This is popularly know as 'outsourcing' and is a matter of considerable contention within the countries from which production is being shifted. It is also being translated notably by countries like the United States, as the last chapter indicated, into claims by national labour movements that the net cost to workers is a real loss of employment opportunities.

It is suggested that, given the increasing growth of industrial production outside the country of origin, firms like Boeing have become 'systems integrators'. This pattern of development has been common to the major MNEs for a number of years. Its major effects are to redistribute the sources of production, such as sub-unit assembly and other forms of construction, across a wide range of firms, some of which are rivals in the market for the final aircraft to be delivered to the client airlines. What this means in theoretical terms is the decline of the conventional model of the industrial firm as a market competitor that keeps its rivals at arm's length at all times.

At the same time, it makes the very important matter of funding, especially for new types as well as existing marques of aircraft, very much subject to major infusions of FDI. This means that while production is distributed across national borders, so is the effective control over the sources of investment. More importantly, while a quid pro quo arrangement may well see production locations decided on the basis of funding by firms in another country, it may also involve the costs of a particular process being carried out in the home country being paid for by funding sourced by FDI.

The perceived threat from what used to be called in some countries the 'runaway factory' now seems to be deepening, as outsourcing is followed up by 'offshoring'. This means that some firms, and a notable example can be found in the airline industry, who have located their IT services offshore at popular sites in countries like India are taking the process a stage further. They are doing so by locating their entire operations offshore as a means of offsetting the relative unit labour costs involved, especially in fields like information technology services.

A counter-argument to the wholesale transfer of jobs case is based on the assumption that developed countries that attempt to sustain traditional forms of manufacturing are simply not addressing the structural shifts in skill needs that comes with technological change. In other words, economic modernization is in decline because of a lack of appropriate investment, not only in science and technology research, but also in education and training.

Defining the Scope, Range and Dualities Implicit in Airport Development

We have seen in earlier discussion that aviation in general, and the international airports in particular, have begun to take on developmental dynamics other than their prescribed roles as service providers. Quite apart from their comparative modal advantages of time and distance accessibility, they are also clearly embedded within the macro processes of economic development and its attendant pressures for urbanization, both metropolitan and perimeter based.

The East Asian region may claim to have a degree of advantage over the aviation industries of the western countries, where sheer physical restraints on available locations, as was seen in Chapter 1, are now facing programmes for further international airport development, with a limited range of choices. In turn, many of these are limited by environmental as much as logistical and service dimensions.

By contrast, the East Asian region tends to have a different order of spatial and developmental problems. These largely stem from the fact that the member states of the entire region, including China, have to face strategic issues that are the consequences of rapid change. These have emerged from the geopolitical processes attendant upon the coming of political independence to many of the countries in the region over the last 50 years.

They have also, in the case of countries such as China, Vietnam, Cambodia, Indonesia and the Philippines, emerged as a consequence of either deliberate changes in the ideological direction of the ruling party in the state or as a consequence of significant political instability over time, with the subsequent emergence of a more pragmatic reformist strategy.

As a consequence, while the West has seen the international airport evolve its range of technical and related services with the advancement of aircraft as well as operational and communications technologies, virtually and in parallel with the long history of aviation as an industry, the states of East Asia have been required to respond to the demands for new international airports to service rapidly growing markets and internal transportation needs, in a relatively much shorter period of time. In doing so, the strategic motivation this is less related to the perceived needs of the airport per se and more to a larger context of development contained in some national development plan.

The need to maintain a balance between specific industrial needs in the international airport sector and the larger demands of national and regional developments aimed at positive economic and resultant social growth is generally recognized. As previous discussion has indicated, attempts at maintaining a balance between national as opposed to regional aspirations remains very high on the agendas of bodies such as ASEAN, APEC, Pacific Asia Free Trade and Economic Development (PAFTAD), and the Pacific Economic Collaboration Council (PECC). Unfortunately, as earlier evidence from Thai research indicated, there is a perceived gap between collective intention and operational reality. The classic driver when push finally comes to shove tends to be the tangible benefits to be obtained by putting national interests and aspirations before those of the other states in the region.

The Strategies and Costs Involved in order to Become a Regional Hub

The search for a viable open skies agreement for the East Asia region is a really a sub-set of a much larger aeronautical equation that stretches around the Asia-Pacific region. ICAO signalled in its 2003 World Conference, that an evolutionary strategy, developed at national level, would service to accommodate the fact that there are leading-edge and trailing-edge countries in the matter of reform, and they should be allowed to proceed at their own pace.

Unfortunately, the ability and determination to maintain significant progress requires that individual countries balance the demands for liberalization, which would allow a competitive market to develop within the own air space, and any loss of material benefits that might then accrue. In other words, open skies does not really fit in with flag carrier and local logistics hub dominance by national firms and interests, even if these are then maintained at the cost of economically efficient management.

From an international airport perspective, the issues presented here have a much larger importance than the maintenance of competitive service functions. If the goal is to be a highly competitive super-hub as part of what appears to be an emergent generation of such strategic locations, the first step, as analysts have pointed out, requires that such a hub be located at the core of a secondary hub system, where FDI, most often provided by foreign carriers, provides the necessary funding in exchange, of course, for contractual rights over slots, services and the strategic advantages that flow from that particular locational choice. This raises the further question of the degree and extent to which the existing national airport system can carry the operational weight of such a role, as well as the very significant cost. This, in turn, raises the need to both obtain and sustain a significant level of investment as an emergent player in international civil aviation.

The Question of Cost and the Strategic Response of China: A Current Example

There are clear indications that East Asia has a strong identity, both as a tourism and business destination, which is increasingly shared by China, especially in the light of the heavy flows of FDI into the national economy. These have been given a specific bias towards the aviation industry, with the decision by China's largest airport company, Capital Airport Holdings (CAH) to float a significant corporate bond issue (US$ 741 million) on the back of a major loan (US$ 625 million) from the European Investment Bank (EIB), with repayment due over 25 years at an interest rate of 3 per cent.

CAH also controls a significant non-core business in the now common form of hotels, restaurants, shops and other services. It controls some 16 airports, of which seven were acquired in 2004, and has interests in a further four. This gives it a current portfolio with an aggregate value of US$ 8.3 billion. Both the scale of the expansion and its relative speed of development are clearly influenced by the fact that Beijing will host the 2008 Olympic Games. The market possibilities have encouraged a number of foreign carriers to service the China market. The symbolic location of United's new booking office in the Garden Hotel in Guangzhou is both a market signal for the future, and a superb choice of location at a world famous hotel and business site. A further indictor of China's confidence in its future as a major global force in civil aviation is the government's avowed intention to have a new airport located in Beijing by 2008.

The Significance of the China–United States Bilateral Agreement

The expected pattern of growth from the most recent expansion of the terms of this agreement clearly favours the United States in both its scale and scope. The advantage for the American airlines really lies in the fact that China's national fleet is well back on the learning curve with regard to international services.

The pattern is repeated with regard to China's miniscule cargo service, as compared with the 1000+ aircraft available to American companies. In effect, China is faced with a service that employs multi-level marketing systems and currently enjoys something of the order of 40 per cent of world volume in both the passenger and cargo sectors.

From an airport perspective, there is some agreement amongst Chinese aviation experts that there is a material benefit from its airport expansion strategy. The system faces strong competition from the key East Asian locations, which have traditionally been the beneficiaries from international flights whose passengers' final destinations were on the mainland. The development of major sites, especially in the coastal regions, coupled with an expected shift to ultra long-haul point-to-point services, will clearly offset the traditional scheduling.

It remains to note, as a rider to previous discussion, that manifest uncertainties now surround these possibilities, as US–China trade becomes an increasingly contentious issue. Given the fact that China's current account surplus is expected to exceed 6 per cent in 2006, the pressure is now increasing for China to allow its currency to appreciate in the conventional way. The situation is compounded by the fact that reluctance to allow national currencies to appreciate is to be found in some East Asian countries.

The overall problem has been compounded by the fact that the IMF has, to date, consistently failed to address the question as international auditor of exchange rate trends. At the current time, there is no special arrangement with the Asian countries to do this, which leaves the problem of rising imbalances possibly subject to unilateral action by the United States. The implications of a trade war are that it would wreak enormous damage on international business and as well as the acknowledged potential for the growth of aviation services in East Asia.

The Emergence of Airline Bilaterals and the Example of Emirates

The recognition of what might be described as an emergent form of competitive entry is reflective of the increasing duality between internal market competition, provided by the presence of foreign airlines at a specific national location, and the bilateral presence of that country's carriers at other national locations. Perhaps the most useful example of a working system that is literally becoming a global model of what might be called competitive congruency is Emirates.

While the home base is the super-hub of Dubai in the United Arab Emirates (UAE), the airline has extended its linkages across the world, with its networks strategically located at selected major hubs. This allows route changes to intersect in a seamless manner. For example, in the South Pacific, having passed through the Australian hubs to Auckland, Emirates then turns towards the East Asian and North Pacific hubs. While Emirates enjoys both operational strength and increasing size, it has been suggested that these advantages are lost if the values of adaptability in the face of changing market circumstances are not constantly maintained.

Quite clearly, the future holds for international airports the same need to be adaptable and responsive to market opportunity as today's successful carriers, such as Lufthansa, who are currently integrating their super-hub locations with their domestic networks of some 12 hubs both in Germany and the region. This gives them strategic control over international, low-cost and regional users and confirms their role as a growing multifunctional player in the aviation industry.

These working examples raise serious questions for the East Asian region. Quite clearly, the ability to progress beyond the current tangle of bilateral agreements, ASAs and the protection of flag carriers will inevitably stand in the way of further development. Returning to the Lufthansa example cited above, it is a legitimate question to ask whether such achievements would have been viable if Europe did not possess an open skies system?

The Special Characteristics of Air Transportation in the Worldwide Context

The notion of globalization carries with it presumptions of sheer size and extent, and this is reflected in the increasing significance of logistics and supply chain issues. As already noted, questions relating to the availability of aviation services are commonly found within the core business plans of industries. Again and from an airline perspective, the business traveller has been and remains a major client within competitive game plans to grow consumer usage. In sum, aviation on a global basis is responsible for the carriage of millions as well as being the mover of very significant cargo consignments on a daily basis.

Recent research has indicated that despite the sheer volume of consumers, the actual global network can be defined as relatively small with regard to the range of origins and destinations. There remain significant areas of the world, for example, central Asia and some parts of Africa, where the level of both national and per capita incomes are of an order that will not sustain a regular and busy schedule of services.

These problems are further compounded by lack of operational infrastructure. On the other hand, further complications arise in existing networks that discriminate against a broad spectrum of services. The primary characteristics of this order of distribution are described in the following categories.

1. Non-stop connections from a given city have a scale free distribution.
2. This also applies to the number of shortest paths taken through a given airport.
3. The airport nodes with the most connections are not always the key hubs in the network.
4. These anomalies arise because the network has a multi-community structure.

Figure 9.1   The characteristics of the worldwide air transport network
Source: Guimera et al., 2005.

The small world property of the results stems from the city-pair short-path relationship between nodes. This is supported by evidence that found that 95 per cent of the nodes have a peripheral location, which sets the almost total majority of their connections within their own inter and intra communities. It is interesting to find that in the ranking of the top 25 cities by the centrality of their location in the worldwide network, the ratings were Singapore (4), Tokyo (9), Hong Kong (12) Manila, (20) and Bangkok (23).

The Importance of Proximate Economic Institutions as a Factor in Hub Status

Quite apart from the range and quality of services available at the appropriate strategic hub, there are a number of other important factors that can influence the decision of firms to invest and even locate in a given major city. Of these, one of the most significant is, of course, the range and operational scale of financial institutions. The ability to generate indigenous capital investment other than through FDI or developmental loans from international agencies has clearly been a strategy target of ASEAN, APEC and PECC for a very long time, and remains so to this day.

Within the East Asian context, Hong Kong, Singapore and Tokyo have both identity and status amongst the world's leading financial markets. Embedded in the spatial context of the financial markets of the region, these three cities constitute the centre of fiscal and money market activity. They are, in turn, surrounded by a periphery of other cites – Bangkok, Shanghai, Shenzhen, Seoul and Taipei – all of whom are seeking to establish the same status as the other three.

In terms of overall institutional development, as previous discussion has indicated, the essentially unilateral commitment by the individual countries to various aspects of market reform vary, sometimes widely, as the example in Figure 9.2 indicates.

APEC is essentially a cooperative forum, which places great weight on consensus as a tool of progress. It gives great weight to freedom and openess as the guiding principles in market reform. These may be encapsulated in the rubric shown in Figure 9.2.

1. Cooperation and interdependence between the countries of the region.
2. The concerted unilateral liberalization of markets.
3. The general raising of living standards throughout the region.
4. The elimination of international trade and investment distortions.
5. Effective competition policies with resultant consumer benefits.

Figure 9.2   The ultimate objective of APEC's policy objectives
Source: P.J. Lloyd & Associates, 2001.

The expected outcomes from these proposals remain in the medium to long term, for, as we have seen, they are subject as much to geopolitical as economic reasoning by the political leadership of the member states. In the interim and somewhat cynically, APEC, it would seem, remains as a 'perfect excuse to chat', and on an annual basis.

Some Comments on Market Competition in East Asia

It is clear at the current time that the East Asian airport industry contains a small elite group of internationally recognized institutions, which have built a solid and growing market on an international basis. It is also clear that there is a significantly larger group whose evolution is either being circumscribed by a lack of developmental equity or low status in some national plan for economic development.

It is this distribution that will probably see the elite group maximize their strategic grip over the immediate term. This is not only a function of their current status, but because they are able to provide a range of changing services as the shifts in market demand dictate. They will be assisted by the fact that those major new sites such as Incheon, Baiyun, Kuala Lumpur and the new Bangkok airport, will be faced with the immediate tasks of gaining not only a market share, but also some degree of return on the enormous amount of capital that has been invested. The situation for some of the new players is further complicated by the fact that as in the case of Incheon, for example, the full development cycle will not be completed until 2020.

A further complication arises in those countries, such as Vietnam, where as already noted earlier, the strategic focus of the central aviation authorities gives a dual emphasis to the role of aviation. There the purpose is perceived as offering both a means of integration within the larger regional and global contexts of international market competition and as a major tool for the development of free population movement within the internal economy.

Some Further Implications of the IT Revolution

The specialized transportation needs of markets that are using supply chain technologies appear to be increasingly met by the major developments of the dedicated cargo firms. DHL, UPS and FedEx are already developing their own hub arrangements and, in doing so, revolutionizing the conventional notions of sheer size and central location as the primary focus of airport development.

Within this specialized form of operational context, the increasing value of information as commercialized goods is also coming to the fore. It is being materially assisted by important new distinctions now appearing within the market. The example of virtual supply centres, which tend to operate to some degree as spot markets and the conventional notion of bulk movement between producer and client come immediately to mind.

Again, the notion that value is found in bulk transmissions, while germane if maritime or road transport are the subject of discussion, is an obvious attempt to maintain a balance between final price to the consumer, delays occasioned by the mode of transport and the costs incurred in real time on voyage. By contrast, air cargo allows a significant advantage to the supplier, since high value combined with low mass can be achieved. In turn, where product discrimination is not a major factor, a virtual market, say for fresh vegetables, can easily match the product and deliver to the customer in real time.

The primary link is, of course, information in the form of essential and usable knowledge, which in some major industries is now the key driver of the business. It is contained in the role of system instigator, which, as earlier discussion has already revealed, has become Boeing's major corporate strategy. An example follows, which reveals the essential flexibility of such systems.

The Case of the Switch Box Company

The firm of Li and Fung of Hong Kong, which has offices in Singapore and other key sites, is an old family company with a major link to the international textile industry. From a core data warehouse, the company is able to manage a wide and varied range of functions from interfacing with leading clothing chains through maximizing production by location all across national boundaries, within the Southeast Asian region.

As a result, the firm is able to maximize efficiencies in real time. This is because its network is based on the famous just in time (JIT) production system pioneered by major Japanese companies. Quite clearly, the ability to move products within designated time frames is reinforced by the availability of dedicated air transportation. This networked pattern of activities, enabled and controlled by the constant flow of knowledge and information, is more than a portent for the future. Forms are already exploring their options in terms of both business-to-business and business-to-customer demand.

The Possibility of Emergent Competition between Specialized and General Carriers

It is interesting to compare this IT-driven example with the more conventional practice followed by mainstream carriers. While conjoint traffic will undoubtedly continue to maintain a share of the international market, the question arises, how much will it retain as firms commit to e-commerce style operations? From a regional point of view, the emergence of dedicated cargo hubs with increasingly sophisticated links to both technology and clients might change the entire configuration of the key export industries within East Asia.

The Strategic Significance of Emergent Sub-national Market Networks

The continued organizational response to the flexibility of movement inherent in the e-commerce and associated business activities is clearly following the movement of international business across national boundaries. The tendency, as found in the models developed by Porter and Ohmae, is to see sub-national locations or business clusters become commercial centres in their own right. Once these clusters reach a critical size as urban conurbations, they may well become independent centres with global network potential.

In effect, this allows them to become semi-independent of their spatial locations. The relative size of such centres is modest when compared with the super metropolitan regions discussed earlier. In effect, they have the power to operate within existing regional areas, but with their developmental directions set by the networks within which they are embedded. The nation state of Singapore is a clear and working example of this type of development. It remains only to comment that the strategic location of airports would have a major role in the successful operation of such a network.

It was Charles Dickens who began one of his most famous novels, *A Tale of Two Cities*, with the immortal words: 'It was the best of times, it was the worse of times.' They are quoted here as a fitting descriptor of the state of the aviation industry at the moment. For while the market signals future growth and aircraft technology continues to maximize user efficiencies on a truly global scale, geopolitical regulative and economic as well as social and cultural imbalances still raise major problems.

From a strategic perspective, there can be little doubt that the search for the undoubted and material benefits that will emerge from the projected growth of East Asia as a global force in world air transport will be led by those economies who can claim, with every justification, to be modern and developed countries. This has been a somewhat speculative search for those primary factors that shape the evolutionary dynamics of the region's aviation industry.

There remains a serious need for a more detailed and analytical examination of some of the questions that have been raised earlier and, with it, a further need for a wide-ranging discourse within the industry. The focus of debate will have to range very widely and, in doing so, will have to take on board the primary fact that the aviation industry has a very broad-based operational and functional role, embedded as it is within the larger developing context of international business and trade.

Addenda

The developments cited are reported on 13 September 2005. It was also confirmed on the same day, that a US$ 1billion investment placed by the major information technology firm Ericsson. See *United States Information Service in China*, 2005.

The research samples do not include cargo and non-passenger flights. The major source of the flight data was OAG and the geographical coordinates of the major airports came from Landings.com. See Guimera et al., 2005.

The APEC notes were taken from P.J. Lloyd & Associates, 2001; see also, Reszats, 2002.

It is important to recall that the applications of JIT, were in fact developed with the external and efficient sectors of the Japanese economy, and supported by its major international corporations. See Porter et al., 2000.

Bibliography

AAPA (2004) *Proceedings of the Annual General Meeting of the Asian Pacific Airlines Association*. Kuala Lumpur, Malaysia, AAPA, p. 20.

Abegglen, J.C. (1994) *Sea Change: Pacific Asia as the New World Industrial Center*. New York, Free Press.

ACI-Europe (2003) 'Models of Contractual Relationships', *Aviation Week and Space Technology*, 161(15): 48.

ACI-Europe (2004) 'World-wide and Regional Forecast: 2005–2020', Annual General Meeting, Geneva.

Air Councils International (2005) *Air Cargo World Annual Report 2005*. Geneva, ACI.

Advani, A. (1999) 'Passenger Friendly Airports: Another Reason for Airport Privatization', Reason Public Policy Institute, Policy Study, 254: 1–25.

Armstrong, D. and Terry, E. (2004) *Pearl River Super Zone: Tapping into the World's Fastest Growing Economy*. Hong Kong, SCMP Book Publishing.

Armstrong, P., Glyn, A. and Harrison, J. (1984) *Capitalism since World War II*. London, Fontana.

Asanjuma, S. (2004) 'The Role of Policy Planning and Coordination in Asia's Infrastructural Development', *Asia Development Bank–JBIC–The World Bank, East Asia and Pacific Infrastructural Development Flagship Study*. Manilla, Asia Development Bank-JBIC-The World Bank, pp. 1–38.

ASEAN (2004) 'Air Transport', *Annual General Meeting Report and Appendices*, Section 13.

Baily, M.N. (2005) 'The United States Economic Outlook?', *Institute of International Economics Briefing Paper*. Washington, DC, Institute of International Economics.

Barrett, S.D. (2000) 'Airport Competition in the Deregulated European Market', *Journal of Air Transport Management*, 6(1): 13–27.

Barzagan, M. and Vasigh, B. (2003) 'Size versus Efficiency: A Case Study of US Commercial Airports', *Journal of Air Transport Management*, 9(3): 187–93.

Baumol, W.J., Panzar, J. and Willig, R. (1987) *Contestable Markets and the Theory of Industry Structure*. San Diego, Harcourt Brace Janovitch.

Beijing International Capital Airport Newsletter (2005) 'East Asian Airport Alliance Reports Progress on Formation', 6 June 2005, pp. 1–4.

Brakman, S., Garnetsen, H. and van Marrij, K.C. (2001) *An Introduction to Geographical Economics: Trade, Location and Growth*. Cheltenham, Edward Elgar.

Bergsten, F.C. (2005a) 'Embedding Pacific Asia in the Asian Pacific', *Institute of International Economics, East Asian Conference*. Tokyo, Japan Press Club.

Bergsten, F.C. (2005b) 'The Trans-Pacific Imbalance: A Disaster in the Making', PECC Annual General Meeting, Seoul.

Bergsten, F.C. and Noland, M. (1993) *Pacific Dynamism and the International Economic System*. Washington, DC, Institute of International Economics.

Berry, S.T., Spiller, P.T. and Carnall, M. (1996) 'Airline Hubs: Costs, Mark-ups and the Implications of Customer Heterogeneity', Working Paper, 5861, National Bureau of Economic Research, Washington, DC.

Bilotkach, V. (2004) 'Asymmetric Regulation and Airport Dominance in International Aviation', University of Arizona School of Management Working Paper Series.

Borestein, S. (1991) 'The Dominant Firm Advantage in Multi-product Industries, Evidence from the US Airline Industry', *Quarterly Journal of Economics*, 106(4): 1237–66.

Borestein, S. (1993) 'Localized Market Power in the US Airline Industry', *The Review of Economics and Statistics*, 75(1): 66–75.

Borestein, S. (1989) 'Hubs, High Fare Dominance, and Market Power in the US Airline Industry', *Rand Journal of Economics*, 20(1): 344–365.

Borestein, S. (1990) 'Airline Mergers, Airport Dominance and Market Power', *American Economic Review, Papers and Proceedings*, 80(2): 400–404.

Botton, N. and McManus, J. (1999) *Competitive Strategies for Service Organizations*, London, Routledge.

Brock, J. (2000) 'Industry Updates: Airlines', *Review of Industrial Organization*, 16: 41–51.

Brooking, A. (1997) 'Studies in Intellectual Capital', *Long Range Planning*, 31(1): 1–29.

Buckley, P.J. and Casson, M. (2002) *The Future of the Multinational Firm, 25th Anniversary Edition*. London, Palgrave Macmillan, originally published 1977.

Buckley, P.J. and Ghuari, P.N. (2004) 'Globalization, Economic Geography and the Strategy of Multinational Enterprises', *Journal of International Business Studies*, 35(1): 81–8.

Burnett, R. (1994) *The Law of International Business Transactions*. Sydney, The Federation Press.

Botton, N. and McManus, J. (1999) *Competitive Strategies for Service Organizations*. London, Routledge.

Calder, S. (2002) *No Frills: The Truth Behind the Low Cost Revolution*. London, Virgin Books.

Canterbury, E.R. (2001) *A Brief History of Economics: An Artful Approach to the Dismal Science*. Singapore, World Scientific.

Carruthers, A., Baipai, J.N. and Hummels, D. (2004) *Trade Logistics: An East Asian Perspective*. New York, The World Bank.

Cartier, C. (2001) *Globalizing South China, Royal Geographical Society with IGB*. Oxford, Blackwell.

Castells, M. (2001) *Challenges of Globalization*. Capetown, Manskew-Miller Longman.

Castells, M. and Hall, P. (1994) *Technopoles of the World: The making of the 21st Century Industrial Complexes*. London, Routledge.

Centre for Asia Pacific Aviation (2005) *Database*. Sydney, Centre for Asia Pacific Aviation.

Christensen, C.N., Anthony, S.D. and Roth, E.A. (2004) *Seeing What's Next?: Using the Theories of Innovation to Predict Industry Change*. Boston, Harvard Business School Press.

Cline, W. (2005) 'US Fiscal Adjustment and External Debt', Institute of International Economics, *New Paper*, p. 5.

Conway, M.K. (1993) *Airport Cities: The New Global Centres of the Twenty First Century*. Norcross, GA, Conway Data Inc.

Davies, R.E.G. (2002) 'Air Transport Directions in the Twenty-first Century', in D. Jenkins (ed.), *Handbook of Airline Economics*, 2nd edition. New York, McGraw-Hill for Aviation Week, pp. 3–26.

De Neufville, R. (1999) *Airport Privatization Issues for the United States*, MIT Technology Policy Working Papers (draft 2), Cambridge Mass., MIT.

Dhume, S. (2002) 'Buying Fast into Southeast Asia', *Far Eastern Economic Review*, 28 March, pp. 30–33.

Dicken, P. (1999) *Global Shift: Transforming the World Economy*, 3rd edition. Liverpool, Paul Chapman.

Doganis, R. (2001) *The Airline Business in the 21ˢᵗ Century*. London, Routledge.

Doganis, R. (2002) *Flying off Course: The Economics of International Airlines*. London, Routledge, 3ʳᵈ edition.

Donne, M. (1995) *The Future of International Air Passenger Transport: A Financial Times Management Report*. London, Pearson Professional.

Douglas, M. (1998) 'East Asian Urbanization, Patterns, Problems and Prospects', Asia-Pacific Research Centre, Stanford University, The Walter Shorestein Distinguished Lecture Series Number One.

Drysdale, P. (2005) 'Regional Cooperation in East Asia and FTA Strategies', Australia–Japan Research Centre, APSDEP-ANU, *Pacific Economic Paper*, 344.

Drysdale, P. and Garnaut, R. (1993) 'The Pacific: Applications of a General Theory of Economic Integration', in F.C. Bergsten and M. Nolan (eds), *Pacific Dynamism and the International Economic System*. Washngton, DC, Institute of International Economics.

Dymond, W.A. and de Mestral, A. (2003) 'New Directions in International Airline Policy', *Policy Matters*, 4(2): 1–15.

Economist, The (2005) 'American Airlines Flying on Empty', 22 September 2005, pp. 57–58.

Elek, A., Findlay, C., Hooper, P. and Warren, T. (1998) 'Open Skies, Open Clubs and Open Regionalism', Australian Productivity Commission, Trade Services Workshop, Canberra.

Emmons, W. (2000) *The Evolving Bargain: Strategic Implications of Deregulation and Privatization*. Boston, Harvard Business School Press.

Ferguson, N. (2001) *The Cash Nexus: Money and Power in the Modern World: 1700–2000*. London, Penguin.

Findlay, C. (1997) 'The APEC Air Transport Schedule', Australia-Japan Research Centre, APSDEP-ANU, *Pacific Economic Paper*, 273.

Findlay, C. and Goldstein, A. (2004) 'Liberalization and Foreign Direct Investment in Asian Transport Systems: The Case of Aviation', *Asian Development Review*, 21(1): 37–65.

Findlay, C. and Pangestu, M. (2001) 'Regional Trade Agreements in Asia-Pacific, Where are They Taking Us?', PECC, Trade Policy Forum, Bangkok.

Findlay, C., Hufbauer, G.F. and Jaggi, G. (1996) 'Aviation in Asia Pacific', in G.F. Hufbauer and G. Jaggi (eds), *Flying High: Liberalizing Civil Aviation in Asia Pacific*. Washington, DC, The Institute of International Economics, pp. 11–32.

Francis, G., Fidato, A. and Humphries, I. (2003) 'Airport–Airline Interaction', *Journal of Air Transport Management*, 9(3): 267–73.

Fridstrom, L.F., Hjelde, F., Lange, H., Murray, E., Norkel, A., Pedersen, T.T., Rytter, N., Talen, C.S., Skoven, M. and Solhaug, L. (2004) 'Towards a More Vigorous Competition Policy in Relation to the Aviation Market', *Journal of Air Transport Management*, 10(1): 3–14.

Fridstrom, L, Hjelde, F., Murray, E., Norkel, A., Pedersen, T.T., Rytter, N., Talen, C.S., Skoven, M. and Solhaug, L. (2003) 'Toward a More Vigorous Market Deregulation', *Journal of Air Transport Management*, 10(1): 71–9.

Friedman, T. (1999) *The Lexus and the Olive Tree*. London, Harper-Collins.

Gillen, D. and Morrison, W.G. (2003) 'Bundling Integration and the Delivery Price of Air Travel: Are Low Cost Carriers Full Service Competitors?', *Journal of Air Transport Management*, 9(1): 15–29.

Gillen, D. and Lall, A. (2004) 'Competitive Advantage of Low Cost Carriers: Some Implications for Airports', *Journal of Air Transport Management*, 10(1): 41–50.

Gordon, D.J., Blaza, A. and Sheate W.R. (2005) 'A Sustainability Risk Analysis of the Low cost Airline Sector', *World Transport Policy and Practice*, 11(1): 13–33.

Gonenc, R. and Nicoletti, G. (2001) 'Regulation, Market Structure and Performance in Air Passenger Transport', *OECD Economics Working Papers*, 7, p. 51.

Graham, A. (2003) *Managing Airports*, 2nd edition. Amsterdam, Elsevier.

Graham, B. (1995) *Geography and Air Transport*. Chichester, Wiley.

Gresham, S.O. and Xu, G. (2004) 'China Moves to Increase Private and International Developments in Airports and Aviation', Morrison Forster Airport Aviation Group, *Legal Updates and News*.

Guimera, R., Mossa, S., Turtschi, A. and Amaral, L.A.N. (2005) 'The World-wide Air Transportation Network: Anomolous Centrality, Community Structure and Cities' Global Roles', *Proceedings of the US National Academy of Science*, 102: 7794–7799.

Gunter, J. (2003) 'Striving to Maintain the Momentum of Change', *ICAO Journal*, 58, 8, 10-13-27-28.

Hanlon, P. (1999) *Global Airlines*. London, Butterworth-Heinemann.

Hasagawa, T. (1996) *Large Scale Airport Construction in Metropolitan Areas: Required Systems and Financial Management*, The Mitsubishi Research Centre, Tokyo, Report 29.

Havel, B.F. (2003) 'A New Approach to the Foreign Ownership of National Airlines', *American Institute of Transport White Paper*. Washington, DC, The American Institute of Transport.

Hayek, F. von (1960) *The Constitution of Liberty*. Chicago, University of Chicago Press.

Heeran, P. and Lamotage, M. (2001) *The Southeast Asian Handbook*. New York, Fitzroy Dearborn.

Hess, S. and Polak, J.W. (2004) 'Development and Applications of a Model for Airport Competition in Multi Airports Regions', Urban Transport Study Group, University of Newcastle on Tyne, UK.

Hesse, M. (2002) 'Missing Links in the Geographies of Distribution', Department of Earth Sciences and Urban Studies, Free University Of Berlin.

Heunmann, R. and Zhang, A. (2002) 'Competition Policy for the Airline Industry in Developing Countries', Conference Report *Pacific Asia Free Trade Area Development (PAFTAD)*, 28, Manilla.

Hill, H. (2000) 'Intra Country Regional Disparities', in *Proceedings of the Second Asian Development Conference*, Singapore, Asia Development Bank ERD Working Series, pp. 1–25.

Holloway, S. (2003) *Changing Planes, A Strategic Management Perspective on an Industry in Transition*, 2. vols. Aldershot, Ashgate.

Hong Kong Aviation Authority (2005) *Press Release: New Arrangements for Services with the Mainland*, 1 September.

Hong Kong Civil Aviation Department (2004) *The Annual Report on Aviation Services*.

Hoyle, B. and Smith, J. (1998) 'Transport and Development of a Conceptual Framework', in B. Hoyle and R. Knowles (eds), *Modern Transport Geography*. Chichester, Wiley, pp. 13–40.

Hufbaeur, G.C. and Findlay, C. (1996) 'Flying High: Liberalization of Civil Aviation in the Asia Pacific', The Institute of International Economics, Washington, DC.

Hutton, W. (2002) *The World We are In*. London, Little Brown.

ICAO (2003a) World Wide Air Transport Conference: Challenges and Opportunities for Liberalization, AT Conf/5-WP/104, Report on Agenda Item 1, Montreal.

ICAO (2003b) World Wide Air Transport AT Conf./5-WP/50 United States Submission on the Liberalization of Market Access, Agenda Items 2.0 and 2.2.

ICAO (2003c) 'Secretariat Review: Airport Congestion, Managing Capacity Restraints', *ICAO Journal*, 55(9): 190–25.

Incheon International Airport Corporation (2001) *The Winged City: Embracing your Dreams with Comfort*. Seoul, Incheon International Airport Corporation, pp. 9–18.

Kahn, A. (1988) 'Surprises of Airline Deregulation', *American Economic Review, Papers and Proceedings*, 78(2) May: 316–322.

Kahn, A.E. (1993) 'The Competitive Consequence of Hub Dominance: A Case Study', *Review of Industrial Organization*, 8(4): 381–405.

Kasarda, J.D. (2000a) *Aeropolis Airports Drive Urban Development in the 21st Century*. New York, The Urban Institute.

Kasarda, J.D. (2000b) *New Logistics Technologies and Infrastructure for the Digitized Economy*. New York, The Urban Institute.

Kasarda, J.D. and Green, J. (2004) 'Air Cargo: Engine for Economic Development', International Air Cargo Association Symposium, (TIACO) Bilbao, Spain.

Kasarda, J.D. and Rondinelli, D.A. (1998) 'Innovative Infrastructures for Agile Manufacturers', *Sloan Management Review*, Winter, 39(2): 73–82.

Keeling, D. (1995) 'Transport and the World City Paradigm', in P. Knox and A. Taylor (eds) *World Cities in a World System*. New York, Cambridge University Press, pp. 115–131.

Keynes, J.M. (1936) *The General Theory of Employment, Interest and Money*. London, Macmillan.

Kogut, B. and Gittelman, M. (2002) 'Globalization', in W. Lazonick (ed.), *The Handbook of Economics, International Encyclopaedia of Business and Management*. London, Thomson, pp. 435–51.

Larsen, A.P. (2000) 'The Future of Air Service Liberalization', *Economic Perspectives*, 5(3): 11–23.

Latter, T. (2004) 'Selling Chek Lap Kok: A Flight of Fancy?', in D. Armstrong and E. Terry (eds), *Pearl River Super Zone: Tapping into the World's Fastest Economy*. Hong Kong, SCMP Publishing, pp. 234–236.

Lau, L.I. (2003) 'Economic Growth in the Digital Era', International Symposium on Welcoming the Challenge of the Digital Era, Taipei.

Leary, M. (2005) 'Address on Industrial Re-regulation', presented at the Conference of Low Cost Carriers, Paris, Air Transport World, 16 June.

Lee, D. (2003) 'An Assessment of Some Recent Criticisms of the US Airline Industry', *Review of Network Economics*, 2(1): 1–8.

Lelieur, M. (2003) *Law and Policy of Substantial Ownership and Effective Control of Airlines*. Aldershot, Ashgate.

Levine, M.E. (2002) 'Price Discrimination without Market Power', *Yale Journal of Regulation*, 19(1): 2–26.

Levine, M.E. (2003) 'Looking Back and Ahead: The Future of the US Airline Industry', MIT Global Aviation Industry Research Programme, Working Paper Series /03.

Loy, F. (1996) 'Questioning the Conventional Wisdom', in G.F. Hufbauer and C. Findlay (eds), *Flying High: Liberalization of Civil Aviation in the Asia-Pacific*. Washington, DC, Institute for International Economics, pp. 117–137.

Lynn, M. (1998) *Birds of Prey: Boeing versus Airbus*. New York, Four Walls and Eight Windows Press.

Mathews, J. (2001) 'Catching Up Strategies in Technology Development, with Particular Reference to East Asia', UNIDO–World Industrial Development Report Background Paper.

Mercer, D. (1999) *Future Revolutions*. London, Orion Books.

Mineta, N. (2005) *US Policy with Regard to Aviation Market Liberalization*, Speech to the American Chamber of Commerce in Hong Kong, 1 June.

Monnell, A. (1992) 'Infrastructure, Investment and Economic Growth', *Journal of Economic Perspectives*, 6: 189–98.

Morrison, W.G. (2004) 'Dimensions of Predatory Pricing in Air Travel Markets', *Journal of Air Transport Management*, 10(1): 87–95.

Mussa, M. (2005) 'Global Economic Prospects, Growth Slowing below Potential', Institute of International Economics, Washington, DC.

Nikomborirak, D. (2001) 'Service Liberalization in ASEAN', *Thailand Development Research Institute Quarterly Review*, 16(4): 12–19.

OECD Secretariat (1997) 'The Future of Air Transport Policy: Responding to Global Change', in *Project on International Air Transport*. Paris, OECD.

O'Sullivan, S. (2000) 'Best Practice for Providing PSP Investment in Infrastructure', Asia Development Bank, Manilla.

Oum, T.H. and Yu, C.Y. (2000) *Shaping Air Transport in Asia-Pacific*. Aldershot, Ashgate.

Paez, A. (2001) 'Network Accessibility and the Spatial Distribution of Economic Activity in East Asia', McMaster University, School of Geography, Centre for Spatial Analysis, Working Paper No. 5.

Panitchpakdi, S. and Clifford, M.L. (2002) *China and the WTO*. Singapore, Wiley.

Park, V.C., Errata, S. and Cheong, I. (2005) 'The Political Economy of the Proliferation of FTAs', *PAFTAD*, 30, Honolulu.

Park, Y. (2003) 'Analysis of the Competitive Strength of Asia's Major Airports', *Journal of Air Transport Management*, 9(3): 353–60.

Park, Y. and Kwon, O.K. (2003) 'Airport Development and Air Cargo Logistics: Korea's Initiative in Northeast Asia', conference paper for *PECC International Round Table: Role of Airports and Airlines in Trade Liberalization and Growth*, Singapore.

Park, Y.C., Urata, S. and Cheong, I. (2004) 'Does Trade deliver on its Promises?', *Pacific Asia Free Trade and Development, 30th Annual Meeting*, Hawaii.

Pels, E. and Verhoef, E.T. (2002) 'Airport Pricing', Free University of Amsterdam, Tinbergen Institute, Discussion Paper, TI 2002-078/3.

Pitelis, C.N. and Schnell, C.A. (2002) 'Barriers to Mobility in Europe's Civil Aviation Markets: Theory and New Evidence', *Review of Industrial Organization*, 20(2): 127–150.

P.J. Lloyd & Associates (2001) *Harmonising Competition and Investment Policies in the East Asian Region*, Third Asia Development Programme, 'Regional Cooperation in Asia and the Pacific'. Manila, Asia Development Bank.

Porter, M.E. (1990) *The Competitive Advantage of Nations*. London, Macmillan.

Porter, M.E. (1998) 'Clusters and the New Economics of Competition', *Harvard Business Review*, November–December: 77–90.

Porter, M.E. (2000) 'Location, Competition, and Economic Development, Local Clusters in the Global Economy', *Economic Development Quarterly*, 14(1): 14–21.

Porter, M.E., Takeuchi, H. and Sakaibara, M. (2000) *Can Japan Compete?* London, Macmillan.

Proceedings of the Third ASEAN Transport Ministers Conference (1997) 'Joint Press Statement 5 June 1997', Cebu, The Phillipines.

Rajan, R.S. and Sen, R. (2004) 'The New Wave of FTAs with Particular Reference to ASEAN, China and India', Asia Development Bank, Workshop on Free trade Areas, Manilla.

Reszats, B. (2002) 'Developing Financial Markets in East Asia-Opportunities and Challenges in the 21st Century', Asia Business Forum on Asset Securitisation, Kuala Lumpur.

Rhoades, D. (2003) *The Evolution of International Aviation, Phoenix Rising.* Aldershot, Ashgate.

Rigg, J. (1997) *Southeast Asia: The Human Landscape of Modernization.* London, Routledge.

Rimmer, P.J. (1994) 'Regional Economic Intergration in the Pacific Area', *Environment and Planning*, 26(11): 1731–1759.

Rodrique, J.P. (1992) 'Transportation and Territorial Development in the Singapore Extended Metropolitan Region', *Singapore Journal of Tropical Geography*, 15(1): 56–74.

Rodrique, J.P. (1996) 'Transportation Corridors in the Asia Pacific Region', in D.A. Hensher and J. King (eds), *Proceedings of the 7th World Conference on Transportation Research.* Sydney, Pergamon Press.

Rugman, A. and Brain, C. (2003) 'Multinational Enterprises are Regional not Global', *The Multinational Business Review*, 11(1): 3–12.

Saitch, A. (2001) *Governance and the Politics of China.* London, Palgrave.

Sandel, M. (1984) *Liberalism and its Critics.* Oxford, Blackwell.

Sander, C. (2002) 'Airport Consolidation: Trends and Opportunities', UNISYS, *Aviation White Paper.*

Sassen, S. (1998a) 'The Impact of New Technologies and Globalization on Cities', in F.C. Lo and Y-M. Yeung (eds), *Globalization of the World's Largest Cities.* Tokyo, UN University Press, pp. 391–409.

Sassen, S. (1998b) *Globalization and Its Discontents.* New York, New Press.

Schott, J. and Watel, J. (2000) 'Decision Making in the WTO', Institute of International Economics, Washington, DC.

Scollay, R. (2000) 'FTA Developments in the Asia-Pacific Region', PECC Trade Forum, Phuket, Thailand.

Scollay, R. (2001) 'RTA Trends in the APEC Region', *Pacific Economic Cooperation Council, Trade Policy Forum*, Bangkok, 12-13 June: 1-17.

Shaafsma, M. (2003) 'Airports and Cities in Networks', DISP Working Paper 154, Schipol Airport.

Shane, J. (2003) 'US Department of Transportation', address to the AIA/IATA Joint Summit and AGM, Washington, DC.

Shapiro, C. and Varian, H.R. (1999) *Information Rules: A Strategic Guide to the Network Economy.* Boston, Harvard Business School Press.

Shen, T. (1992) *Reform and Opening in China's Civil Aviation Industry*. Beijing, International and Cultural Press.

Sideri, S. (1997) 'Globalization and Regional Integration', *European Journal of Developmental Research*, 9(1): 81–8.

Silva, G.F. (1999) 'Public Policy for the Public Sector', The World Bank Group, Note 202, pp. 1–2.

Singer, P. (2002) *One World: The Ethics of Globalization*. Melbourne, Text Press.

Sit, V. (2004) 'Global Transpark: New Competitiveness for Hong Kong and South China based on Logistics', *Geographikar Annaler*, 86B: 145–163.

Slane, J. (2003) *Keynote Speech to the International Air Transport Association AGM Summit Meeting*, Washington, DC, 1-3 June.

Smith, D. and Timberlake, M. (2002) 'Global Cites and Globalization in East Asia: Empirical Realities and Conceptual Questions', University of California at Irvine, Center for the Study of Democracy Issues, 02/09.

Soros, G. (1998) *The Crisis of Global Capitalism*. New York, Little Brown.

Stiglitz, J. (2003) *The Roaring Nineties: Seeds of Destruction*. London, Penguin Books.

Thant, M., Tang, M. and Kakazu, H. (1994) (eds) *Growth Triangles in Asia: A New Approach to Regional Economic Cooperation*. Hong Kong, Asia Development Bank and Oxford University Press.

Thomas, G. (2004) 'Fighting for a Place at the Table', *Air Transport World*, December 2004, 1.

Turnbull, P. (1995) 'Regulation, Deregulation or Reregulation of Transport', ILO Symposium on the Social and Labour Consequences of the Deregulation and Privatization of Transport, Discussion Paper No.4, Geneva.

UN (2000) 'Department of Economic and Social Affairs, World Urbanization Prospects', New York.

United States Information Service in China (2005), 13 September 2005, internet source.

US Department of State (2005) *Bilateral Aviation Agreement with Thailand: Media Note 9 September 2005*, Agreement ratified 19 September 2005.

Van Londen, S. and de Ruitjer, A. (2003) *Managing Diversity in a Glocalizing World*, Lavona, Italy, Fondazione Eni Enrico Matei.

Vasigh, B. and Haririan, M. (2003) 'An Empirical Investigation of Financial and Operational Efficiency of Private versus Public Airports', *Journal of Air Transportation*, 8(1): 91–109.

Vernon-Wortzel, H. and Wortzel, L.H. (1999) *Strategic Management in a Global Economy*. Chichester, John Wiley.

Webster, D. (2002) 'On the Edge: Shaping the Future of Peri-urban East Asia', Stanford Pacific Asia Research Centre, Working Paper 1–53.

Weisbrod, G.E., Reed, J.S. and Neuwirth, R.E. (1993) 'An Airport Area Development Model', in *Proceedings of the PTRC International Transport Conference*. Manchester, PTRC.

Welch, D.E. and Worms, V. (2005) 'International Business Travellers, a Challenge for Human Resource Management', in G.K. Stahl and I. Borkman (eds), *Handbook of Research in International Human Resource Management*. Cheltenham, Edward Elgar, pp. 1–12.

Wells, A.T. and Wensveen, J.G. (2004) *Air Transportation: A Management Perspective*, 5th edition. Australia, Thomson.

Wikipedia.com (undated) *Low cost Airlines in Asia*, wikipedia.com.

Williams, A. and Williams, B.A. (2005) 'Structural Change in the International Airline Industry: Major Human Resource Issues now facing the US Legacy Carriers', in *Proceedings of the 7th World Conference in International Human Resource Management*, Cairns, Australia, 14–18 June, pp. 171–184.

Williams, G. (1994) *The Airline Industry and the Impact of Deregulation*, revised edition. Aldershot, Avebury Aviation.

Yergin, D. and Stanislaw, J. (1998) *The Commanding Heights: The Battle Between Government and the Marketplace that is Remaking the Modern World*. New York, Simon and Schuster.

Yoshikazu, T. (2000) 'Preparation of East Asia Passenger O-D Tables and Dynamic Analysis: Report on Research Trends, and Accomplishments', Japan, CAA of Japan.

Zhang, A. (1997) 'Industrial Reform and Air Transport Management in China', University of Victoria, Department of Economics, Occasional Paper No.17.

Index